Carbon Fiber Composites

Carbon Fiber Composites

Deborah D. L. Chung

Butterworth–Heinemann

Boston   London   Oxford   Singapore   Sydney   Toronto   Wellington

Library of Congress Cataloging-in-Publication Data

Chung, Deborah D. L.
    Carbon fiber composites / Deborah D. L. Chung.
        p.   cm.
    Includes bibliographical references and index.
    ISBN 978-0-75069-169-7
    1. Fibrous composites.   2. Carbon fibers.   I. Title.
TA418.9.C6C533 1994
620.1'93—dc20   94-11427
                                              CIP

British Library Cataloguing-in-Publication Data
A catalogue record for this book is available from the British Library.

Butterworth–Heinemann
313 Washington Street
Newton, MA 02158

Transferred to digital printing 2005

**To the memory of my maternal grandmother,
Lee Sun Chau, M.D. (1890–1979)**

Chau with the author

Contents

PART **II**

Carbon Fiber Composites

Preface

Carbon fiber composites, particularly those with polymeric matrices, have become the dominant advanced composite materials for aerospace, automobile, sporting goods, and other applications due to their high strength, high modulus, low density, and reasonable cost. For applications requiring high temperature resistance, as required by spacecraft, carbon fiber carbon-matrix composites (or carbon–carbon composites) have become dominant. As the price of carbon fibers decreases, their applications have even broadened to the construction industry, which uses carbon fibers to reinforce concrete. An objective of this book is to provide up-to-date information on the whole spectrum of carbon fiber composites, including polymer-matrix, metal-matrix, carbon-matrix, ceramic-matrix, and hybrid composites. Such information pertains to the processing, properties, and applications, and is given in a tutorial fashion, so that no prior knowledge of the field is required. At the end of each chapter, a large number of up-to-date references are included, so the reader can look up further information, if desired. Thus, the book is suitable for students as well as professionals.

In contrast to other books on composites, this book is focused on composites with a variety of matrices but carbon fibers alone as the filler. This focus allows detailed consideration of the fiber–matrix interface and the composite processing for a variety of matrices. In contrast, most books on composites tend to focus on only polymeric matrices and consider a variety of fillers. Because of the rapidly increasing importance of matrices other than polymers, the new focus used in this book is necessary.

This book was written from the viewpoint of materials, whereas most books on composites were written from the viewpoint of mechanics. Although mechanics is important for the design of structural composite components, the development of new advanced composite materials depends more heavily on composite processing and the fiber–matrix interface. Because this book addresses a variety of emerging composite materials, an emphasis on processing rather than mechanics is necessary. Furthermore, composite materials are increasingly used for applications other than structure; an example is electronic packaging, which uses composites for the sake of their electrical and thermal properties mainly. Because of the emphasis on materials rather than mechanics, this book is more suitable for materials scientists/engineers than mechanical or aerospace engineers.

In addition to covering carbon fiber composites (Part II), this book also covers carbon fibers (Part I). The reader does not have to read Part I before Part II. However, a basic understanding of carbon fibers helps one appreciate the use of the fibers in various matrices. In keeping with an emphasis on materials, Part I is included in this book.

This book is suitable for use as a textbook in the senior undergraduate and graduate levels. Basic undergraduate chemistry and materials science are the only subjects needed for effective use of the book. It is most suitable for courses in materials departments, chemistry departments, and chemical engineering departments. Although this book is unconventional compared to other books on composites, it fills a partial vacuum in the area of composites from the viewpoint of materials rather than mechanics and in the area of composites with matrices other than polymers. I appreciate the value of the other books on composites, and I hope this book will complement them in serving students and professionals in the field of composite materials.

Carbon Fiber Composites

Carbon Fibers

CHAPTER 1

Introduction to Carbon Fibers

Carbon fibers refer to fibers which are at least 92 wt.% carbon in composition [1]. They can be short or continuous; their structure can be crystalline, amorphous, or partly crystalline. The crystalline form has the crystal structure of graphite (Figure 1.1), which consists of sp^2 hybridized carbon atoms arranged two-dimensionally in a honeycomb structure in the x–y plane. Carbon atoms within a layer are bonded by (1) covalent bonds provided by the overlap of the sp^2 hybridized orbitals, and (2) metallic bonding provided by the delocalization of the p_z orbitals, i.e., the π electrons. This delocalization makes graphite a good electrical conductor and a good thermal conductor in the x–y plane. The bonding between the layers is van der Waals bonding, so the carbon layers can easily slide with respect to one another; graphite is an electrical insulator and a thermal insulator perpendicular to the layers. Due to the difference between the in-plane and out-of-plane bonding, graphite has a high modulus of elasticity parallel to the plane and a low modulus perpendicular to the plane. Thus, graphite is highly anisotropic.

The high modulus of a carbon fiber stems from the fact that the carbon layers, though not necessarily flat, tend to be parallel to the fiber axis. This crystallographic preferred orientation is known as a fiber texture. As a result, a carbon fiber has a higher modulus parallel to the fiber axis than perpendicular to the fiber axis. Similarly, the electrical and thermal conductivities are higher along the fiber axis, and the coefficient of thermal expansion is lower along the fiber axis.

The greater the degree of alignment of the carbon layers parallel to the fiber axis, i.e., the stronger the fiber texture, the greater the c-axis crystallite size (L_c), the density, the carbon content, and the fiber's tensile modulus, electrical conductivity, and thermal conductivity parallel to the fiber axis; the smaller the fiber's coefficient of thermal expansion and internal shear strength.

The carbon layers in graphite are stacked in an AB sequence, such that half of the carbon atoms have atoms directly above and below them in adjacent layers (Figure 1.1). Note that this AB sequence differs from that in a hexagonal close packed (HCP) crystal structure. In a carbon fiber, there can be

3

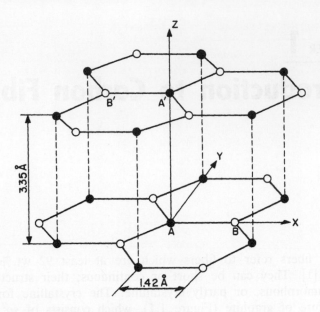

Figure 1.1 The crystal structure of graphite.

graphite regions of size L_c perpendicular to the layers and size L_a parallel to the layers. There can also be crystalline regions in which the carbon layers, though well developed and parallel to one another, are not stacked in any particular sequence; the carbon in these regions is said to be turbostratic carbon. Yet another type of carbon that can exist in carbon fibers is amorphous carbon, in which the carbon layers, though well developed, are not even parallel to one another.

The proportion of graphite in a carbon fiber can range from 0 to 100%. When the proportion is high, the fiber is said to be graphitic, and it is called a graphite fiber. However, a graphite fiber is polycrystalline, whereas a graphite whisker is a single crystal with the carbon layer rolled up like a scroll. Because of their single crystal nature, graphite whiskers are virtually flaw-free and have exceptionally high strength. However, the production yield of graphite whiskers is too low for them to be commercially significant. This book only deals with fibers—not whiskers.

Table 1.1 [2] compares the mechanical properties, melting temperature, and density of carbon fibers with other types of fibers. There are numerous grades of carbon fibers; Table 1.1 only shows the two high-performance grades, which are labeled "high strength" and "high modulus." Among the fibers (not counting the whiskers), high-strength carbon fibers exhibit the highest strength while high modulus carbon fibers exhibit the highest modulus of elasticity. Moreover, the density of carbon fibers is quite low, making the specific modulus (modulus/density ratio) of high-modulus carbon fibers exceptionally high. The polymer fibers, such as polyethylene and Kevlar fibers, have

Table 1.1 Properties of various fibers and whiskers.

Material	Density[a] (g/cm³)	Tensile strength[a] (GPa)	Modulus of elasticity[a] (GPa)	Ductility (%)	Melting temp.[a] (°C)	Specific modulus[a] (10^6 m)	Specific strength[a] (10^4 m)
E-glass	2.55	3.4	72.4	4.7	<1 725	2.90	14
S-glass	2.50	4.5	86.9	5.2	<1 725	3.56	18
SiO_2	2.19	5.9	72.4	8.1	1 728	3.38	27.4
Al_2O_3	3.95	2.1	380	0.55	2 015	9.86	5.3
ZrO_2	4.84	2.1	340	0.62	2 677	7.26	4.3
Carbon (high-strength)	1.50	5.7	280	2.0	3 700	18.8	19
Carbon (high-modulus)	1.50	1.9	530	0.36	3 700	36.3	13
BN	1.90	1.4	90	1.6	2 730	4.78	7.4
Boron	2.36	3.4	380	0.89	2 030	16.4	12
B_4C	2.36	2.3	480	0.48	2 450	20.9	9.9
SiC	4.09	2.1	480	0.44	2 700	12.0	5.1
TiB_2	4.48	0.10	510	0.02	2 980	11.6	0.3
Be	1.83	1.28	300	0.4	1 277	19.7	7.1
W	19.4	4.0	410	0.98	3 410	2.2	2
Polyethylene	0.97	2.59	120	2.2	147	12.4	27.4
Kevlar	1.44	4.5	120	3.8	500	8.81	25.7
Al_2O_3 whiskers	3.96	21	430	4.9	1 982	11.0	53.3
BeO whiskers	2.85	13	340	3.8	2 550	12.3	47.0
B_4C whiskers	2.52	14	480	2.9	2 450	19.5	56.1
SiC whiskers	3.18	21	480	4.4	2 700	15.4	66.5
Si_3N_4 whiskers	3.18	14	380	3.7	—	12.1	44.5
Graphite whiskers	1.66	21	703	3.0	3 700	43	128
Cr whiskers	7.2	8.90	240	3.7	1 890	3.40	12

[a]From Ref. 2.

densities even lower than carbon fibers, but their melting temperatures are low. The ceramic fibers, such as glass, SiO_2, Al_2O_3 and SiC fibers, have densities higher than carbon fibers; most of them (except glass fibers) suffer from high prices or are not readily available in a continuous fiber form. The tensile stress–strain curves of the fibers are straight lines all the way to fracture, so the strength divided by the modulus gives the ductility (strain at break) of each fiber, as shown in Table 1.1. The main drawback of the mechanical properties of carbon fibers is in the low ductility, which is lower than those of glass, SiO_2, and Kevlar fibers. The ductility of high-modulus carbon fibers is even lower than that of high-strength carbon fibers.

Carbon fibers that are commercially available are divided into three categories, namely general-purpose (GP), high-performance (HP), and acti-

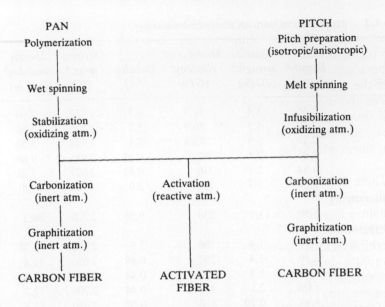

Scheme 1.1   The processes for making carbon fibers from PAN and pitch precursors.

vated carbon fibers (ACF). The general-purpose type is characterized by an amorphous and isotropic structure, low tensile strength, low tensile modulus, and low cost. The high-performance type is characterized by relatively high strength and modulus. Among the high-performance carbon fibers, a higher modulus is associated with a higher proportion of graphite and more anisotropy. Activated carbon fibers are characterized by the presence of a large number of open micropores, which act as adsorption sites. The adsorption capacity of activated carbon fibers is comparable to that of activated carbons, but the fiber shape of activated carbon fibers allows the adsorbate to get to the adsorption site faster, thus accelerating the adsorption and desorption processes [3]. The amount adsorbed increases with the severity of activation. Severe activation may be achieved by treating commercial ACF with sulfuric acid followed by heating at up to 500°C [4].

Commercial carbon fibers are fabricated by using pitch or polyacrylonitrile (PAN) as the precursor. The processes for both precursors are shown in Scheme 1.1 [5].

Precursor fibers are fabricated by conventional spinning techniques, such as wet spinning for PAN and melt spinning for pitch. They must be converted to a form which is flameproof and stable at the high temperatures (> 700°C) involved in carbonization. Therefore, before carbonization (pyrolysis), they are stabilized for the case of the PAN precursor, or infusiblized for the case of the pitch precursor. Both stabilization and infusiblization are carried out in an oxidizing atmosphere. After that, general-purpose and high-performance fibers are obtained by carbonization in an inert atmosphere, followed by graphitization at > 2 500°C in an inert atmosphere if a high modulus is desired, whereas activated carbon fibers are obtained by activating in a reactive atmosphere, such as steam at elevated temperatures. To enhance the preferred orientation in the high-performance carbon fibers, graphitization can be performed while the fibers are under tension. The higher the graphitization temperature, the greater the preferred orientation.

For the case of pitch as the precursor, isotropic pitch gives an isotropic carbon fiber, which belongs to the category of general-purpose carbon fibers, whereas anisotropic pitch (such as mesophase pitch) gives high-performance carbon fibers which have the carbon layers preferentially parallel to the fiber axis.

Table 1.2 shows the tensile properties of various carbon fibers on the market. Among the high-performance (HP) carbon fibers, those based on pitch can attain a higher modulus than those based on PAN, because pitch is more graphitizable than PAN. In particular, the HP fiber designated E-130 by du Pont exhibits a modulus of 894 GPa, which is over 80% of the theoretical value of a graphite single crystal (1 000 GPa). A higher modulus is associated with a lower elongation at break, as shown by comparing the group of Amoco HP fibers in the order P-25, P-75S, and P-120S, and the group of du Pont HP fibers in the order E-35, E-75, and E-130. The du Pont HP fibers exhibit higher tensile strengths and greater elongations than the Amoco HP fibers of similar moduli. Among the high-performance (HP) fibers, those based on PAN can attain a higher tensile strength and greater elongation than those based on pitch, because (1) shear is easier between the carbon layers in a graphitized fiber, (2) pitch is more graphitizable than PAN, and (3) the oriented graphitic structure causes the fibers to be more sensitive to surface defects and structural flaws. In particular, the HP PAN-based fiber designated T-1000 by Toray exhibits a tensile strength of 7 060 MPa and an elongation of 2.4%. The general-purpose (GP) fibers tend to be low in strength and modulus, but high in elongation at break.

Table 1.2 also shows the diameters of various commercial carbon fibers. Among the HP fibers, those based on PAN have smaller diameters than those based on pitch.

Pitch-based carbon fibers (GP and HP) represent only about 10% of the total carbon fibers produced worldwide around 1990 [6], but this percentage is increasing due to the lower cost and higher carbon content of pitch compared to PAN. The costs of precursors and carbon fibers are shown in Table 1.3. Mesophase pitch-based carbon fibers are currently the most expensive, due to the processing cost. Isotropic pitch-based carbon fibers are the least expensive. PAN-based carbon fibers are intermediate in cost.

Figure 1.2 [7] shows the prices and tensile strengths of carbon fibers, aramid fibers, and glass fibers. Although carbon fibers are mostly more expensive than aramid fibers or glass fibers, they mostly provide higher tensile strengths. Among the different grades of carbon fibers, the prices differ greatly. In general, the greater the tensile strength, the higher the price.

The price of carbon fibers has been decreasing, while the consumption

Table 1.2 Tensile properties and diameters of commercial carbon fibers.

Type	Fiber designation	Tensile strength (MPa)	Tensile modulus of elasticity (GPa)	Elongation at break (%)	Diameter (μm)	Manufacturer
GP	T-101S	720	32	2.2	14.5	Kureha Chem.
	T-201S	690	30	2.1	14.5	Kureha Chem.
	S-210	784	39	2.0	13	Donac
	P-400	690	48	1.4	10	Ashland Petroleum
	GF-20	980	98	1.0	7–11	Nippon Carbon
HP (PAN)	T-300	3 530	230	1.5	7.0	Toray
	T-400H	4 410	250	1.8	7.0	Toray
	T-800H	5 590	294	1.9	5.2	Toray
	T-1000	7 060	294	2.4	5.3	Toray
	MR 50	5 490	294	1.9	5	Mitsubishi Rayon
	MRE 50	5 490	323	1.7	6	Mitsubishi Rayon
	HMS-40	3 430	392	0.87	6.2	Toho Rayon
	HMS-40X	4 700	392	1.20	4.7	Toho Rayon
	HMS-60X	3 820	588	0.65	4.0	Toho Rayon
	AS-1	3 105	228	1.32	8	Hercules
	AS-2	2 760	228	1.2	8	Hercules
	AS-4	3 795	235	1.53	8	Hercules
	AS-6	4 140	242	1.65	5	Hercules
	IM-6	4 382	276	1.50	5	Hercules
	HMS4	2 484	338	0.7	8	Hercules
	HMU	2 760	380	0.70	8	Hercules
HP (pitch)	P-25	1 400	160	0.9	11	Amoco
	P-75S	2 100	520	0.4	10	Amoco
	P-120S	2 200	827	0.27	10	Amoco
	E-35	2 800	241	1.03	9.6	du Pont
	E-75	3 100	516	0.56	9.4	du Pont
	E-130	3 900	894	0.55	9.2	du Pont
	F-140	1 800	140	1.3	10	Donac
	F-600	3 000	600	0.52	9	Donac
ACF	FX-100	—	500[a]	18[b]	15	Toho Rayon
	FX-600	—	1 500[a]	50[b]	7	Toho Rayon
	A-10	245	1 000[a]	20[c]	14	Donac
	A-20	98	2 000[a]	45[c]	11	Donac

[a]Specific surface area (m^2/g).
[b]Adsorption amount of benzene (%).
[c]Adsorption amount of acetone (%).

Table 1.3 Cost of PAN-based, mesophase pitch-based, and isotropic pitch-based carbon fibers. From Ref. 6.

	Cost of precursor ($/kg)	Cost of carbon fibers ($/kg)
PAN-based	0.40	60
Mesophase pitch-based	0.25	90
Isotropic pitch-based	0.25	22

has been increasing, as shown in Figure 1.3 [7]. The decreasing price is broadening the applications of carbon fibers from military to civil applications, from aerospace to automobile applications, and from biomedical devices to concrete structures.

Under rapid development are short carbon fibers grown from the vapor of low-molecular-weight hydrocarbon compounds, such as acetylene. This process involves catalytic growth using solid catalyst particles (e.g., Fe) to form carbon filaments, which can be as small as 0.1 μm in diameter. Subsequent chemical vapor deposition from the carbonaceous gas in the same chamber causes the filaments to grow in diameter, thus resulting in vapor grown carbon fibers (VGCF) or gas-phase grown carbon fibers.

Carbon fibers can alternatively be classified on the basis of their tensile strength and modulus. The nomenclature given below was formulated by IUPAC.

- UHM (ultra high modulus) type: carbon fibers with modulus greater than 500 GPa

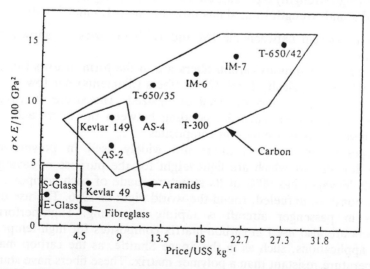

Figure 1.2 The product of tensile strength (σ) and tensile modulus (E) versus price for various commercial carbon fibers. From Ref. 7. (By permission of Pion, London.)

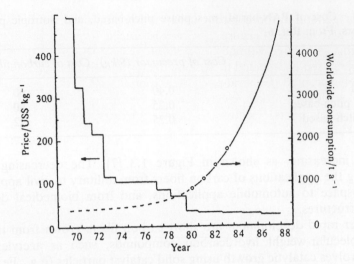

Figure 1.3 Changes in carbon fiber price and consumption over the last 20 years. From Ref. 7. (By permission of Pion, London.)

- HM (high modulus) type: carbon fibers with modulus greater than 300 GPa and strength-to-modulus ratio less than 1%
- IM (intermediate modulus) type: carbon fibers with modulus up to 300 GPa and strength-to-modulus ratio above 1×10^{-2}
- Low-modulus type: carbon fibers with modulus as low as 100 GPa and low strength. They have an isotropic structure
- HT (high strength) type: carbon fibers with strength greater than 3 GPa and strength-to-modulus ratio between 1.5 and 2×10^{-2}

There is overlap between the IM and HT categories, as shown by the above definitions.

Commercial continuous carbon fibers are in the form of tows (untwisted bundles) containing typically 1 000–12 000 fibers (filaments) per tow, or yarns (twisted bundles). They may be sized or unsized. The sizing improves the handleability and may enhance the bonding between the fibers and certain matrices when the fibers are used in composites.

High-performance carbon fibers are widely used in polymer-matrix composites for aircraft which are lightweight for the purpose of saving fuel. The aircraft Voyager has 90% of its structure made of such composites and achieved a nonstop, unfueled, round-the-world flight in 1986. The use of such composites in passenger aircraft is rapidly increasing. High-performance carbon fibers are also used in carbon-matrix composites for high-temperature aerospace applications, such as in the Space Shuttle, as the carbon matrix is more temperature resistant than a polymer matrix. These fibers have started to be used in metal matrices, such as aluminum, for aerospace applications, as aluminum is more temperature resistant than polymers.

Short general-purpose pitch-based carbon fibers are used for the reinforcement of concrete, because low cost is crucial for the concrete industry. Because this is a large-volume application of carbon fibers, the tonnage of carbon fibers used is expected to increase markedly as this application becomes more widely accepted. General-purpose carbon fibers are also used for thermal insulation, sealing materials, electrically conducting materials, antistatic materials, heating elements, electrodes, filters, friction materials, sorbents, and catalysts [8].

References

1. E. Fitzer, in *Carbon Fibers Filaments and Composites*, edited by J.L. Figueiredo, C.A. Bernardo, R.T.K. Baker, and K.J. Huttinger, Kluwer Academic, Dordrecht, 1990, pp. 3–41.
2. D.R. Askeland, *The Science and Engineering of Materials*, 2d ed. PWS-Kent, 1989, p. 591.
3. L.I. Fridman and S.F. Grebennikov, *Khimicheskie Volokna* 6, 10–13 (1990).
4. Isao Mochida and Shizuo Kawano, *Ind. Eng. Chem. Res.* 30(10), 2322–2327 (1991).
5. K. Okuda, *Trans. Mater. Res. Soc. Jpn.* 1, 119–139 (1990).
6. D.D. Edie, in *Carbon Fibers Filaments and Composites*, edited by J.L. Figueiredo, C.A. Bernardo, R.T.K. Baker, and K.J. Huttinger, Kluwer Academic, Dordrecht, 1990, pp. 43–72.
7. E. Fitzer and F. Kunkele, *High Temp.–High Pressures* 22(3), 239–266 (1990).
8. R.M. Levit, *Khimicheskie Volokna* 6, 16–18 (1990).

CHAPTER **2**

Processing of Carbon Fibers

Introduction

Carbon fibers are fabricated from pitch fibers, polymer fibers (e.g., polyacrylonitrile), or carbonaceous gases (e.g., acetylene). Those made from pitch and polymer fibers are commercially available, whereas those made from carbonaceous gases are not yet commercially available. Those made from pitch and polymer fibers are in short and continuous forms, whereas those made from carbonaceous gases are in the short form only. Fibers made from pitch and carbonaceous gases are more graphitizable than those made from polymers, so they can attain higher thermal conductivity and lower electrical resistivity. The raw materials cost is much lower for making fibers from pitch or carbonaceous gases than from polymers. However, the present market is dominated by those made from polymers because of their combination of good mechanical properties (particularly tensile strength) and reasonable cost. In contrast, highly graphitic pitch-based carbon fibers are very expensive, though they have high tensile modulus, high thermal conductivity, and low electrical resistivity. On the other hand, the price of pitch-based carbon fibers is expected to drop when the production volume increases. Eventually, the least expensive carbon fibers are expected to be those made from pitch and carbonaceous gases.

The fabrication of carbon fibers from pitch or polymers involves pyrolysis of the pitch or polymer; this is performed by heating. Pyrolysis is akin to charring, thus forming carbon. In contrast, the fabrication of carbon fibers from carbonaceous gases involves catalytic growth of carbon. For the pyrolysis process, pitch has the advantage of having a higher carbon yield than polymers.

This chapter describes the methods for fabricating carbon fibers from all three types of precursors. The chemical treatments and coatings needed to enhance the bonding with various matrices in composite materials are addressed in the chapters on composite materials. As carbon fibers are most conveniently used in a woven form, weaving is also covered in this chapter.

Carbon Fibers made from Pitch

Pitch used as a precursor for carbon fibers can be a petroleum pitch (such as a distillation residue obtained by the distillation of crude oil under atmospheric or reduced pressure, or a heat-treated product of the by-product tar obtained by the pyrolysis of naphtha), a coal tar pitch, or other pitches. Coal pitches are, in general, more aromatic than petroleum pitches. Table 2.1 [1] shows a chemical analysis of typical commercial pitches derived from petroleum and coal. The coal pitch has a much higher benzene and quinoline-insoluble content. A high quinoline-insoluble content usually means that the material has a high solid content. These solid carbon particles can accelerate coke formation during subsequent thermal processing of the pitch and lead to fiber breakage during extrusion and thermal treatment. Therefore, although petroleum pitches are less aromatic, they are more attractive as precursors for carbon fibers [2].

Pitch is a thermoplast, so it melts upon heating. The melt can be spun to form pitch fibers. The pitch fibers must be pyrolyzed (carbonized) by heating at ≥ 1 000°C to form carbon fibers and they must maintain their shape during carbonization, so they must first undergo infusiblization (stabilization). Infusiblization is a process for rendering the pitch infusible. This process involves air oxidation at 250–400°C. After carbonization at ≥ 1 000°C in an inert atmosphere, graphitization is optionally carried out at ≥ 2 500°C, if a high modulus, a high thermal conductivity, or a low electrical resistivity is desired. The higher the graphitization temperature, the more graphitic is the resulting fiber. High-strength HT-type carbon fibers are formed after carbonization whereas high-modulus HM-type carbon fibers are formed after graphitization. If isotropic pitch is used as the precursor, the graphitization heat treatment has to be carried out while the fiber is being stretched. This costly process, called stretch-graphitization, helps to improve the preferred orientation in the fiber. On the other hand, if anisotropic pitch is used as the precursor, stretching is not necessary, because the anisotropic pitch has an inherently preferred orientation of its molecules.

Isotropic pitch can be converted to anisotropic pitch by heating at 350–450°C for a number of hours [3,4]. The anisotropy refers to optical anisotropy; the optically anisotropic parts shine brightly if the pitch is polished and observed through the crossed nicols of a reflection type polarized light microscope. The anisotropy is due to the presence of a liquid crystalline phase, which is called the mesophase. The mesophase is in the form of small liquid droplets. Within each droplet, large planar molecules line up to form nematic order. The droplets (spherules) grow in size, coalesce into larger spheres and eventually form extended anisotropic regions. The so-called mesophase pitch is a heterogeneous mixture of an isotropic pitch and the mesophase. The relative amounts of the two phases can be determined approximately by extraction with pyridine or quinoline. The isotropic fraction is soluble in pyridine, while the

Table 2.1   Properties of coal and petroleum pitches. From Ref. 1.

	Coal pitch	Petroleum pitch[a]
Softening point (°C)	116	120
Glass transition temperature (°C)	34	48
Mass % C	91.9	93.3
Mass % H	4.13	5.63
Mass % S	0.73	1.00
Mass % N	1.17	0.11
Mass % O	1.05	0.37
Percent aromatic H (NMR)	82	60
Percent benzene insoluble	32	7
M_n (benzene soluble)	460	560
Percent quinoline insoluble	12	0.2

[a]Ashland 240 (a petroleum pitch).

mesophase is insoluble due to its high molecular weight. The mesophase has a higher surface tension than the low-molecular-weight isotropic liquid phase from which it grows. As the proportion of mesophase increases, the viscosity of the pitch increases, so a higher temperature is required for subsequent spinning of the pitch into fibers. As the mesophase is heated, its molecular weight increases (it polymerizes), cross-linking occurs and the liquid eventually becomes solid coke. This solidification must be avoided in the spinning of carbon fibers. Moreover, the difference in density between the two phases in mesophase pitch causes sedimentation of the mesophase spherules. Although the sedimentation can be decreased by agitating the pitch, it makes it difficult to obtain homogeneous fibers from the pitch. Hence, there are pros and cons about the presence of the mesophase. Nevertheless, mesophase pitch is important for producing high-performance pitch-based carbon fibers.

A method to produce mesophase pitch involves (1) heating the feed pitch (e.g., at 400°C in N_2 for 14–32 h. [5]), with or without de-ashing and distillation, to transform to mesophase pitch containing 70–80% mesophase, (2) allowing to stand at a slightly lower temperature so that the mesophase sinks, and (3) separating the mesophase by centrifuging [4,6]. Mesophase pitch is used in the UCC process [5], now owned by Amoco.

Due to the shortcomings of mesophase pitch, neomesophase pitch (optically anisotropic), dormant anisotropic pitch (optically isotropic) and premesophase pitch (optically isotropic), are used [3].

The neomesophase pitch is produced by first removing the high-molecular-weight component by solvent extraction (using an aromatic solvent such as toluene), as this component tends to form coke upon heating, and then heating at 230–400°C [3]. The neomesophase has a lower softening temperature than the mesophase, so it can be spun at a lower temperature, which reduces the coke formation [4]. Neomesophase pitch is used in the EXXON process [7], which is now owned by du Pont.

The dormant anisotropic pitch is between isotropic and mesophase pitches in nature. It is dormant in the sense that it does not interfere with spinning, but, on heating after spinning, it becomes active and orients itself [4]. The process of forming dormant anisotropic pitch involves (1) heating pitch at 380–450°C to form anisotropic pitch containing several percent of mesophase, (2) hydrogenation of the anisotropic pitch to form isotropic pitch with a lower softening temperature, and (3) heating the isotropic pitch at 350–380°C to form dormant anisotropic pitch [1]. This pitch results in a carbon fiber between GP and HP grades and with high elongation [6].

The premesophase pitch is formed by (1) hydrogenation at 380–500°C using hydrogen donor solvents (such as tetrahydroquinoline), H_2/catalysis and other techniques, and (2) heating the hydrogenated pitch at $> 450°C$ for a short time. This pitch is optically isotropic at the spinning temperature, but orients readily during heating subsequent to spinning. Coal-derived pitch is preferred to petroleum-derived pitch for this method, which is called the Kyukoshi method [3,6].

Dormant anisotropic pitch and premesophase pitch have a rather naphthenic nature, which is caused by hydrogenation [3]. A process not involving hydrogenation but involving the polymerization of naphthalene has been reported for producing an optically anisotropic pitch [8].

The preparation of mesophase pitch from isotropic pitch involves heat treatment at 350–450°C. During the heating, an inert gas such as nitrogen is often bubbled through the pitch to agitate the fluid and to remove the low-molecular-weight components. However, retention of some of these components is vital for the mesophase to have a low quinoline-insoluble (QI) content and a low melting point. For retaining some of these components, a prior heat treatment either in the presence of a reflux or under a moderate pressure is effective [9].

The conversion from isotropic pitch to mesophase pitch is a time-consuming process which can take as long as 44 h. To promote this transformation, an oxidative component can be added to the inert sparging gas. For example, the modified sparging gas can be nitrogen containing 0.1–2 vol.% oxygen [10].

Other than having a high degree of aromaticity, pitches for making carbon fibers should have 88–93 wt.% C, 7–5 wt.% H, and other elements (e.g., S and N) totalling below 4 wt.% [10]. Therefore, feed pitch needs to undergo de-ashing and distillation prior to the various processes mentioned above.

Figure 2.1 summarizes the typical preparation methods of precursor pitch for high-performance carbon fibers.

The spinning of mesophase pitch to form pitch fibers is difficult for a number of reasons [4].

1. The mesophase is viscous.
2. The higher spinning temperature of mesophase pitch compared to isotropic pitch causes additional polycondensation, which leads to gas evolution. Thus, the spinneret needs to be vented to avoid entrapping the gas bubbles in the carbon fibers.
3. The mesophase pitch has a heterogeneous structure, which consists of anisotropic mesophase and isotropic regions.

In spite of the difficulties mentioned above, mesophase pitch is used to produce high-modulus, high-strength carbon fibers having a highly oriented structure [11]. The spinning of mesophase pitch is performed by a variety of conventional spinning methods, such as centrifugal spinning, jet spinning and melt spinning. Melt spinning (more accurately termed melt extrusion) is most commonly used. It involves extruding the melted pitch into a gaseous atmosphere (e.g., N_2) through nozzles directed downward, so that the extruded fibers are cooled and solidified. The melt spinning process is illustrated in Figure 2.2 [2]. An extruder is typically used to melt the pitch and pump it to the spin pack, which contains a filter for removing solid particles from the melt. After passing through the filter, the melt exits the bottom of the spin pack through a spinneret, which is a plate containing a large number of parallel capillaries. An airflow is often directed at the melt exiting from these capillaries in order to cool the fibers. The solidified fiber is finally wound onto a spinning spool. The capillary or orifice has a typical diameter of 0.1–0.4 mm. For spinning 10 μm diameter fibers at 2.5 m/s from an orifice of diameter 0.3 mm, a draw ratio of about 1 000:1 is required. Typical spinning parameters are shown in Table 2.2.

The temperature of the nozzles is determined depending on the type of the pitch and the melt viscosity most suitable for spinning. An increase in the spinning temperature decreases the viscosity of the pitch, as shown in Figure 2.3 [6] for premesophase pitch prepared by the Kyukoshi method and spun at

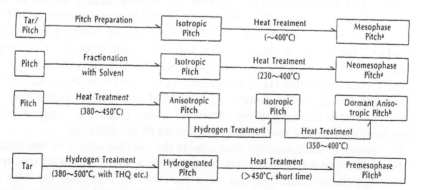

Figure 2.1   Typical preparation methods of precursor pitch for high-performance carbon fibers. From Ref. 3.

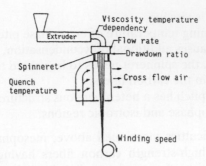

Figure 2.2 Apparatus for the melt spinning of pitch. From Ref. 2. (Reprinted by permission of Kluwer Academic Publishers.)

Table 2.2 Spinning parameters used by two companies. From Ref. 4

	Kureha	*Union Carbide*
Spinning temperature (°C)	240–330	350–436
Spinneret diameter (mm)	0.1–0.9	0.07–0.38
Number of holes	30–120	41–1 000
Spinneret capillary length-to- diameter ratio	n.a.	2–5
Filament diameter (μm)	8–30	8–50
Filament speed (m/min.)	400–1 700	29–226
Draw ratio	400–5 000	3–1 702

n.a. = not available.

300–400°C. The viscosity in turn controls the microstructure of the resulting fiber, also shown in Figure 2.3. When the spinning temperature is 349°C or below, a radial-type structure forms. When the spinning temperature is raised above 349°C, the structure changes from the radial-type structure to either the random-type structure or the radial-type structure surrounded by the onion-skin-type (concentric-circle-type) structure. At very low spinning temperatures, the structure is often accompanied by V-shaped grooves or cracks extending from the circumference toward the center of a fiber (Figure 2.3, rightmost), so it is not desirable for the mechanical properties of the fiber [6].

The design of the spinneret also affects the microstructure of the resulting fiber. Since the radial-type structure results from a laminar flow of mesophase pitch through the spinneret, it can be suppressed by changing the flow from laminar to turbulent. A turbulent flow can be obtained by using a spinneret hole with narrower and wider parts (Figure 2.4b) [6,12] or a hole containing a filter layer of stainless steel particles or mesh (Figure 2.4c) [6].

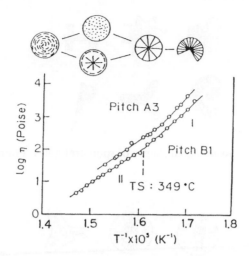

Figure 2.3 Viscosity of mesophase pitches prepared by the hydrogenation method (Kyukoshi method), with schematic microstructures of carbon fibers obtained by spinning at 300–400°C. From Ref. 6.

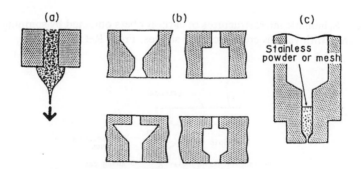

Figure 2.4 Schematic structures of spinnerets that are (a) conventional, or (b) and (c) designed to avoid the formation of a radial-type carbon fiber microstructure. From Ref. 6.

The cross-sectional shape of the spinneret hole can be used to control the microstructure and cross-sectional shape of the resulting fiber, as illustrated in Figures 2.5 and 2.6 [6]. Noncircular carbon fibers are attractive in their increased surface area to volume ratio and the presence of crevices in some of the shapes. The crevices result in increased fiber wetting through capillary action. Thus, noncircular carbon fibers may provide improved fiber–matrix bonding in composites [13].

Short pitch fibers are made by melt blowing rather than melt spinning. As illustrated in Figure 2.7, melt blowing involves melting the pitch and then extruding the pitch through a spinneret, such that a gas stream is passed

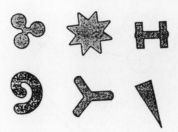

Figure 2.5   Various cross-sectional shapes of spinneret holes. From Ref. 6.

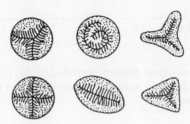

Figure 2.6   Schematic microstructures of carbon fibers obtained by spinning pitch using spinnerets with various cross-sectional shapes. From Ref. 6.

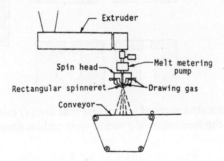

Figure 2.7   Apparatus for the melt blowing of pitch. From Ref. 2. (Reprinted by permission of Kluwer Academic Publishers.)

through the spin head and allowed to exit parallel to the extruding fibers. The spin head heats the gas, and this hot gas draws the fibers down as they emerge from the spinneret capillary. The fibers are blown onto a moving conveyor. No winding takes place. This process is mostly used for producing short isotropic pitch-based carbon fibers [2].

Before carbonization, as-spun pitch fibers are infusiblized or stabilized in order to prevent softening and resulting deformation of the pitch fibers upon heating. Stabilization (also called thermosetting) involves oxidation of the pitch

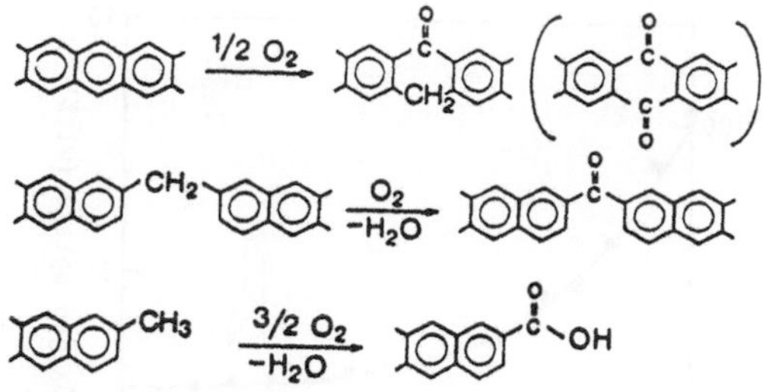

Figure 2.8 Possible reaction mechanisms for pitch oxidation. From Ref. 14. (By permission of *Erdoel und Kohle, Erdgas, Petrochemie.*)

molecules; the intermolecular interactions result in a higher softening point. The heating during stabilization is performed in air at 250–350°C, as provided by a cylindrical furnace placed below the spinning nozzles prior to windup or by blowing hot air through a spool of as-spun pitch fibers. Possible reaction mechanisms for pitch oxidation are shown in Figure 2.8. The direct oxygen attack of individual pitch molecules results in the formation of ketone, carbonyl, and carboxyl groups [14]. The oxidation reaction is accelerated by methyl- and hydro- groups, which also react with carbonyl groups [15]. The introduction of polar CO groups leads to hydrogen bonding between adjacent molecules. During subsequent carbonization at ~1 000°C, the oxidized molecules may serve as starting points for three-dimensional cross-linking [14].

A pitch with a higher softening point permits oxidation to take place at a higher temperature, thereby greatly reducing the time required for oxidation. Therefore, in isotropic pitch fiber production, a pitch with a higher softening point is preferred, even though it is less spinnable [2].

Mesophase pitch fibers have higher softening points than isotropic pitch fibers, so they can undergo stabilization at a higher temperature, which makes the process faster. Indeed minutes are required for stabilizing mesophase pitch fibers, whereas hours are required for stabilizing isotropic pitch fibers [4].

A skin–core structure with the skin richer in oxygen than the core can be introduced in mesophase pitch fibers by incomplete oxidative stabilization. The longer the time of stabilization, the thicker the skin; the skin is the stabilized part. However, stabilization slows down as it proceeds to the interior of a fiber. Figure 2.9 shows the oxygen distribution in a mesophase pitch fiber (30 μm diameter) after oxidative stabilization at 300°C for 15 min. This oxygen distribution corresponds to a skin thickness of 5 μm. After 30 min. of stabilization, the skin thickness has increased to 9 μm; after 90 min. of stabilization, the skin has fully grown, leaving behind no core [16,17].

Incomplete oxidative stabilization may be followed by solvent extraction to help avoid adhesion among the fibers after carbonization. The solvent, such as tetrahydrofuran (THF) or benzene, serves to remove the soluble or fusible fractions in the surface layer. Incomplete oxidative stabilization followed by solvent extraction is called two-step stabilization [18,19].

Due to the low tensile strength of pitch fibers (isotropic or mesophase

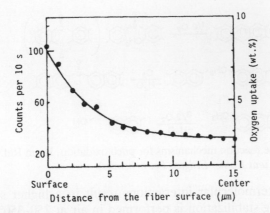

Figure 2.9 Oxygen distribution along the diameter from the surface to the center of a 30 μm diameter stabilized mesophase pitch fiber. From Ref. 16. (Reprinted by courtesy of Chapman & Hall, London.)

pitch), fiber handling should be minimized during stabilization. The tensile strength and modulus of the as-spun mesophase pitch fiber are much lower than those of the carbonized mesophase pitch fiber, as shown in Table 2.3 [2].

In order to reduce fiber sticking and fusion during the stabilization treatment, colloidal graphite, an aqueous suspension of carbon black in ammonium 2-ethylhexyl sulfate, silicone, or other lubricants, can be applied on the surface of the pitch fibers before stabilization [4,20]. The separability of the individual filaments can be improved by using a suspension comprising a silicone oil (e.g., dimethylpolysiloxane) and fine solid particles (e.g., graphite, carbon black, silica, calcium carbonate, etc.), of particle size 0.05–3 μm preferably) [20].

After stabilization, the pitch fibers are carbonized by heating in a series of heating zones at successively higher temperatures ranging from 700–2 000°C. An inert atmosphere is used to prevent oxidation of the resulting carbon fibers. During carbonization, the remaining heteroatoms are eliminated. This is accompanied by the evolution of volatiles, so a gradated series of heating zones is needed to avoid excessive disruption of the structure. The greatest quantity of gases, mainly CH_4 and H_2, are evolved below 1 000°C. Above 1 000°C, hydrogen is the principal gas evolved [2]. For temperatures $\leqslant 2\,000$°C, nitrogen

Table 2.3 Average mechanical properties of mesophase fibers at various stages of processing. From Ref. 2.

	Tensile strength (GPa)	Tensile modulus (GPa)
As-spun mesophase fiber	0.04	4.7
Carbonized mesophase fiber	2.06	215.8

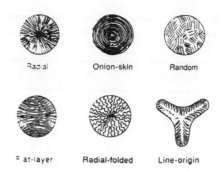

Radial   Onion-skin   Random

Flat-layer   Radial-folded   Line-origin

Figure 2.10   Cross-sectional microstructures of mesophase carbon fibers. From Ref. 2. (Reprinted by permisssion of Kluwer Academic Publishers.)

can be used for the inert atmosphere. For temperatures $>2\,000°C$, nitrogen is not suitable because of the danger of nitrogen reacting with carbon to form a cyanogen; argon can be used instead.

The transverse (cross-sectional) fiber microstructure developed during fiber formation is retained after carbonization. This microstructure is influenced by the flow profile during extrusion from the spinneret and by the fiber elongation prior to windup [2]. Figure 2.10 illustrates the various transverse microstructures in mesophase pitch-based carbon fibers. The lines within each panel of Figure 2.10, whether straight or curved, depict the carbon layers, which are parallel to the fiber axis, at least preferentially.

After carbonization, graphitization is carried out (optionally) by heating in an inert atmosphere at $2\,500–3\,000°C$.

A variation of the melt spinning method involves spinning the pitch upward, such that the molten pitch goes through the spinneret upward and the extrusion face of the spinneret is in contact with a liquid (175–450°C) which has a density greater than that of the pitch. The density difference causes the fiber to move upward (due to buoyancy). On top of and in contact with this liquid layer is another layer of the same liquid at a higher temperature (500–650°C), which causes dehydrogenation of the fiber. The higher-temperature liquid is less dense than the lower-temperature liquid below it, so it stays on top. The liquid may be LiCl or KCl. On top of the top liquid layer is inert atmosphere at an even higher temperature (above 900°C) for further dehydrogenation, which results in carbon fibers containing more than 95 wt.% C. The main attraction of this form of melt spinning is the elimination of the oxidation step due to the support of the weak spun fiber by the high-density liquid around it [21].

Carbon Fibers made from Polymers

Carbon fibers are most commonly made from polymer precursors in the form of textile fibers that leave a residue of carbon and do not melt on

Table 2.4 Weight loss on heating precursor fibers at 1 000°C in helium. From Ref. 22.

Precursor fibers	Weight loss (%)
Pitch	30
PAN preoxidized	38
PAN	60, 67
Saran	74
Rayon	88
Ramie	91

pyrolysis in an inert atmosphere. The polymers include rayon cellulose, polyvinylidene chloride, polyvinyl alcohol, and, most commonly, polyacrylonitrile (PAN). Table 2.4 [22] shows the weight loss on heating pitch and polymer precursor fibers in helium at 1 000°C. The weight loss of PAN fibers is higher than that of pitch fibers, but lower than those of other polymer fibers. In particular, the carbon yield of PAN is about double that of rayon, although PAN fibers are more expensive than rayon fibers. Moreover, PAN fibers have a higher degree of molecular orientation than rayon fibers.

This section focuses on carbon fibers made from PAN. Acrylonitrile, $CH_2=CH-CN$, is the monomer; it has a highly polar nitrile group.

$$CH_2=CH-C\equiv N \quad {}^+CH_2-CH=C=N^-$$

It is polymerized by addition polymerization to PAN.

$$\ldots CH_2 \qquad CH_2 \qquad CH_2 \qquad CH_2 \ldots$$
$$\quad CH \qquad CH \qquad CH$$
$$\quad CN \qquad CN \qquad CN$$

The polymerization can yield a precipitated polymer by using a solvent in which the polymer is soluble. Suitable solvents include dimethyl formamide, dimethyl sulfoxide, and concentrated aqueous solutions of zinc chloride and sodium thiocyanate. All are liquids with highly polar molecular structures, as the polar groups attach to the nitrile groups, thereby breaking the dipole–dipole bonds [23]. The initiators used for the addition polymerization can be the usual ones, such as peroxides, persulfates, azo compounds such as azo-bis-isobutyronitrile, and redox systems [23]. The initiators provide free radicals for the initiation, which is the addition of a radical to an acrylonitrile molecule to form a larger radical.

$$R^\cdot + CH_2{=}CH \rightarrow R{-}CH_2{-}CH^\cdot$$
$$\quad\quad\quad\;\; | \quad\quad\quad\quad\quad\quad |$$
$$\quad\quad\quad\; CN \quad\quad\quad\quad\quad CN$$

where $R^\cdot$ represents a radical.

PAN is a white solid with a glass transition temperature of about 80°C and a melting temperature of about 350°C. However, PAN degrades on heating prior to melting.

Polymer fibers can be fabricated by various spinning methods.

1. Melt spinning: extruding a melt of the polymer.
2. Melt assisted spinning: extruding a homogeneous single-phase melt in the form of a concentrated polymer–solvent blend.
3. Dry spinning (Figure 2.11a [24]): extruding a solution of the polymer in a volatile organic solvent into a circulating hot gas environment in which the solvent evaporates.
4. Wet spinning (Figure 2.11b [24]): extruding a solution of the polymer in an organic or inorganic liquid into a coagulating liquid (a mixture of a solvent and a nonsolvent); this precipitates the polymer, which is then drawn out as a fiber.
5. Dry-jet wet spinning: extruding a solution of the polymer into an air gap (~ 10 mm), followed by a coagulating bath, in order to enhance orientation prior to coagulation.

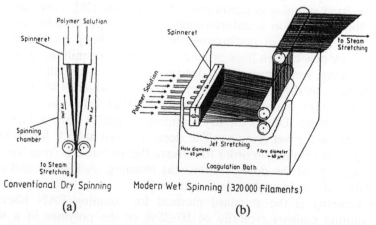

Figure 2.11 Schematic of the polymer spinning process: (a) dry spinning (2 500 filaments) and (b) modern wet spinning (320 000 filaments). From Ref. 24. (By permission of Elsevier Science Publishers B.V.)

All methods involve pumping the melt or solution of the polymer through a large number of small holes in a stainless steel disc, called a spinneret, such that the hole diameter is about twice the final diameter of the fiber. Spinning in clean-room conditions produces better PAN fibers [22].

Because PAN decomposes below its melting temperature, melt spinning is not possible. Melt assisted spinning of PAN uses a solvent in the form of a hydrating agent to decrease the melting point and the melting energy of PAN by decoupling nitrile–nitrile association through the hydration of pendant nitrile groups. With a low melting point, the polymer can be melted without much degradation [25,26]. Water is most commonly used as the hydrating agent. Water-soluble polyethylene glycol (PEG) can also be used [22]. The PAN/water system (1 part PAN to 3 parts water) forms a single-phase solution above 180°C and the solution phase-separates with solidification of the polymer on cooling to 130°C [27]. Carbon fibers with satisfactory mechanical properties have been obtained by extruding at 140–190°C a homogeneous melt consisting essentially of (1) an acrylic polymer containing at least 85 wt.% of recurring acrylonitrile units, (2) approximately 3–20 wt.% of C_1 to C_2 nitroalkane based upon the polymer, (3) approximately 0–13 wt.% of C_1 to C_4 monohydroxy alkanol based upon the polymer, and (4) approximately 12–28 wt.% of water based upon the polymer [25,26]. However, PAN fibers made by melt assisted spinning contain more internal voids and surface defects than those made by wet or dry spinning [22]. On the other hand, PAN fibers made by melt assisted spinning can have a larger variety of cross-sectional shapes:—trilobal, multilobal, for example. Such shapes provide a greater surface area, which enhances fiber–matrix bonding in composites. Because melt assisted spinning of PAN does not require potentially harmful solvents, solvent recovery is not needed and wastewater treatment is not critical, in contrast to the harmful solvents required for dry spinning, wet spinning, and dry-jet wet spinning. Because of this, melt assisted spinning is technologically attractive [28]. However, its use in carbon fiber production is unfortunately not being pursued because of the enormous cost involved in performing the U.S. military qualification tests required for any new carbon fiber product to be used for military purposes.

Dry spinning, wet spinning, and dry-jet wet spinning are all referred to as solution spinning, as they all use a polymer solution, which is known as dope. The dope is not 100% liquid; its solid content (7–30 wt.%) is used to adjust the viscosity of the dope. Dopes for dry spinning generally have a higher solid content than those for wet spinning. The dope is stored at 25°C for about 24 h. before the start of spinning in order to remove the air bubbles from the viscous dope. This storage of the dope is known as ripening. A higher solid content tends to reduce the void content in the fibers [22].

Wet spinning is the standard method for spinning PAN fibers. The spinning solution consists typically of 10–25% of the polymer in a solvent, which can be a mixture of dimethyl formamide and water, a mixture of dimethyl sulfoxide and water, or others. The molecular weight of PAN is in the range 70 000–200 000 and is chosen to yield a solution viscosity that provides a compromise between fiber drawability and final fiber properties. A coherent spinline is formed by phase separation in a suitable coagulating medium, which contains a mixture of a solvent (the same as used for the preparation of dope) and a nonsolvent (water most commonly). The higher the concentration of the nonsolvent, the higher the coagulation rate. The higher the temperature of the coagulation bath, the faster the coagulation. A lower coagulation rate is preferred because a higher coagulation rate causes surface irregularities, greater pore density, and the formation of a skin–core structure. The residence time in the bath is around 10 sec. By using a low concentration of the nonsolvent and a low temperature, PAN fibers in a gel state (i.e., the state prior to coagulation of the extruded dope) can be obtained. The molecular chains in the gel can be quite easily oriented upon stretching because the trapped solvent decreases the cohesive forces among the nitrile groups of the polymer chains. To provide sufficient time to stretch the gel fiber [22], coagulation is slowed down by allowing the gel fiber to pass through several baths containing varying compositions of the coagulation mixture. The coagulated fibers are called protofibers. They are stretched about 2.5 times in the coagulation bath. After washing, a further stretch of about 14 times in steam at 100°C aligns the molecular chains along the fiber axis [4]. The greater the stretching temperature, the greater the draw ratio that can be attained. Similar stretching is applied to PAN fibers fabricated by dry spinning after evaporation of the solvent.

Dry-jet wet spinning is replacing wet spinning because it yields fibers of better mechanical properties and controlled noncircular cross section. Moreover, the spinning speed is higher and the dope can be spun at a higher temperature, so that dopes of higher solid contents can be used. PAN fibers made by dry-jet wet spinning have superior mechanical properties to those made by dry spinning [22].

A tow of a large number of filaments ($\sim$ 50 000 or more) can be produced by wet spinning but not by dry or dry-jet wet spinning. However, fibers of controlled noncircular cross sections can be obtained by dry and dry-jet wet spinning, whereas the cross-sectional shape of fibers made by wet spinning depends on the collapse during coagulation. A lower polymer concentration or a lower temperature in the spinning solution leads to more collapse of the coagulating fiber and more away-from-round shape (even though the spinneret orifices are round) [29]. The greater surface area, resulting from the noncircular cross section, provides better heat flux during stabilization and carbonization; this reduces chain scission and weight loss to produce superior tensile strength and modulus in the resulting carbon fibers [22].

The drawability of a homopolymer (100%) PAN is limited because of the hydrogen bonds in the structure. Therefore, 5–10 mol% of a comonomer is typically added. Examples of comonomers are shown in Table 2.5 [22]. The comonomer is a more bulky acrylic monomer which diminishes the crystallinity of the PAN structure, thereby acting as an internal plasticizer and improving the drawability. Hence, commercial PAN fibers are copolymers.

Table 2.5 Various comonomers for acrylic precursors. From Ref. 22.

Comonomer	Chemical structures
Acrylic acid (AA)	CH_2=CHCOOH
Methacrylic acid (MAA)	CH_2=C(CH_3)COOH
Itaconic acid (IA)	CH_2=C(COOH)CH_2COOH
Methacrylate (MA)	CH_2=CHCOOHCH_3
Acrylamide (AM)	CH_2=CHCONH_2
Aminoethyl-2-methyl propenoate (quaternary ammonia salt)	CH_2=CH(CH_3)COOC_2H_4NH_2

Table 2.6 Effects of comonomers on PAN-based carbon fibers. From Ref. 30.

Comonomer	Property improvement
Allyl sulfonate (0.1–2%)	Tensile strength and modulus improved
Vinyl bromide (4%)	Stabilization time shortened
Carboxyl (5 mol%) containing monomer	Improved tensile strength and good adhesion to matrix
Methyl acrylate (2%), methylene butadiene dioic acid (3%)	Decreases solution viscosity
Comonomers (1.5%) with isopropyl or *tert*-butyl esters of polymerizable unsaturated acids	Good mechanical properties

A second comonomer is often added to initiate ladder-polymer formation (cyclization reaction) during subsequent stabilization of the PAN fibers. It is an acidic comonomer, such as acrylic acid and itaconic acid at concentration levels of about 1 mol% [27]. Among the comonomers in Table 2.5, itaconic acid is particularly effective in helping cyclization because its two carboxylic groups increase the possibility of interaction with the nitrile group, in spite of the dipole–dipole repulsion between the carboxylic and nitrile groups [22].

A third comonomer with basic or acidic pendant groups may be added to make dyeing easier and more controllable. For example, vinyl pyridine is used for acid dyes, and sulfonic vinyl benzene and acrylic acids are used for basic dyes. The amount used is about 0.4–1.0 mol% [23].

Table 2.6 [30] lists the functions of some specific comonomers.

General requirements of the polymer are the following [22]:

- high molecular weight (~ 10^5)
- a molecular weight distribution corresponding to a polydispersity ratio of 2–3 M_w/M_n
- minimum molecular defects

General requirements of the precursor fibers are the following [22]:

- a diameter of 10–12 μm
- high strength and modulus
- a broad exothermic peak due to nitrile group oligomerization during heating and it should start at a low temperature
- a high carbon yield ($>50\%$)

A small diameter of the precursor fiber is desirable for dissipating heat during conversion of the precursor fiber to a carbon fiber, since the heat evolved during the exothermic oligomerization reaction may lead to a low carbon yield. For controlling the heat flux, the rate and initiation temperature of the exothermic reaction should be lowered [22]. The use of a microporous precursor fiber (as obtained by heating with water at 100°C or higher) also helps the conversion to a carbon fiber [31].

The spun PAN fibers typically have a diameter of 11–19 μm, a tensile modulus of 8 GPa, and a tensile strength of 0.5 GPa [27]. (The ratio of the modulus of the carbon fiber to that of the precursor fiber is about 20 [22].) The higher the draw ratio, the greater the modulus and strength, as shown by Figure 2.12. The tensile stress–strain curve shows an initial elastic region, which could be due to the resistance of the CN−H bonds, followed by a regime of plastic flow of increasing resistance to stress until fracture occurs at about 30% elongation [23].

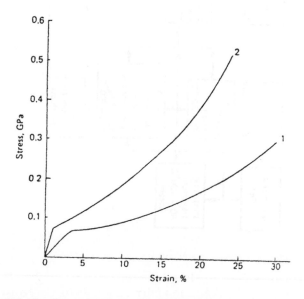

Figure 2.12   Tensile stress–strain curves for PAN fibers obtained at (1) a low draw ratio and (2) a high draw ratio. From Ref. 22.

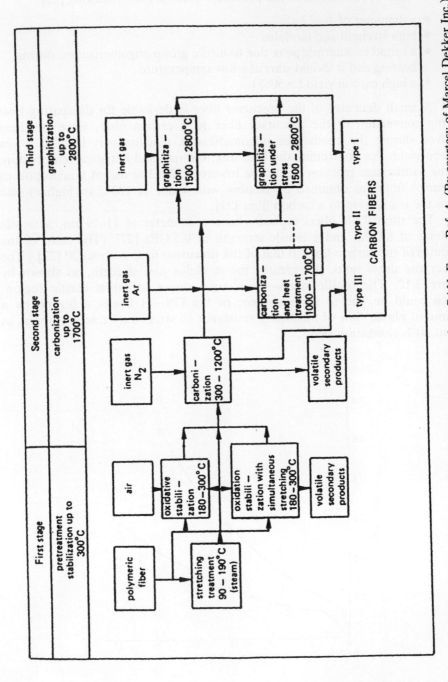

Figure 2.13 Schematic representation of carbon fiber fabrication from PAN. From Ref. 4. (By courtesy of Marcel Dekker Inc.)

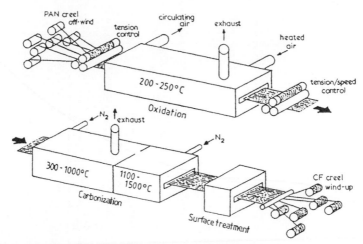

Figure 2.14 Apparatus for the fabrication of carbon fibers from PAN. From Ref. 29. (Reprinted by courtesy of Elsevier Science Publishers B.V.)

A surface finish oil is applied to the spun PAN fibers to assist in handling. The oils are usually volatile above 130°C, so they are removed during subsequent stabilization of the PAN fibers [29]. Other than silicone oil, fatty acid derivatives and guar gum can be used [22].

The conversion of a PAN fiber to a carbon fiber involves stabilization and carbonization. To further increase the modulus, graphitization can be carried out after carbonization. After carbonization (and optionally graphitization), the fibers are given a surface treatment. To produce continuous carbon fibers these steps are performed in a continuous sequence along a production line, as illustrated in Figure 2.13 [29] and Figure 2.14 [4].

Figure 2.14 shows PAN fibers in the form of tows brought off bobbins into a collimated array to pass through the first stage, which is stabilization. Stabilization involves oxidation in air at 180–300°C (preferably below 270°C) under controlled tension and speed. The tension is applied to prevent shrinkage or even cause elongation of the fiber; PAN fibers, when fully relaxed by heating, shrink by about 25% due to the formation of nitrile conjugation cross-links between the polymer chains [23]. During stabilization, gases (NH_3, HCN, etc.) are evolved, so the temperature is controlled by heated air circulation.

The stabilization serves to increase the carbon yield during subsequent carbonization at 300–1 500°C. It converts the thermoplastic PAN into a nonplastic cyclic compound that can withstand the high temperatures during carbonization. The cyclized structure is called a ladder polymer. The conversion is:

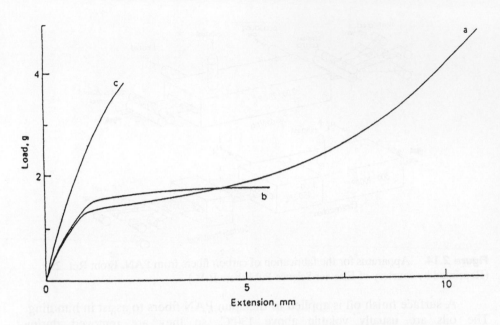

Figure 2.15 Tensile load–extension curves for PAN fibers (a) before and (b) after oxidation at 220°C, and (c) after pyrolysis to 400°C. From Ref. 23. (Reprinted by courtesy of Elsevier Publishers B.V.)

The cyclization initiates through a radical mechanism in the case of the PAN homopolymer and an ionic mechanism in the presence of acid comonomers [22]. It occurs during heating under tension in an inert or oxidizing atmosphere. An oxidizing atmosphere is used because it results in a higher rate of cyclization, a higher carbon yield after subsequent carbonization, and improved mechanical properties of the carbon fibers. Hence, the process is called thermooxidative stabilization. In addition to cyclization, stabilization results in dehydrogenation and three-dimensional cross-linking of the parallel molecule chains by oxygen bonds; the cross-links keep the chains straight and parallel to the fiber axis, even after the release of tension post stabilization. However, the post-stabilization cross-linking is not extensive as shown by the low secondary modulus, indicative of easy plastic flow (Figure 2.15).

Oxygen acts in two opposite ways during stabilization. On the one hand it initiates the formation of activated centers for cyclization, while on the other hand it retards the reactions by increasing the activation energy. In spite of this, oxygen is desirable because it results in the formation of some oxygen-containing groups (such as $-OH, >C=O, -COOH$) in the backbone of a ladder polymer. These groups subsequently help in fusion of the ladder chains during carbonization [22]. Due to the cyclization, the fiber density increases along with the oxygen content during stabilization, as shown in Figure 2.16 [32]. An oxygen content of 8–12 wt.% is present in fully stabilized fibers [32].

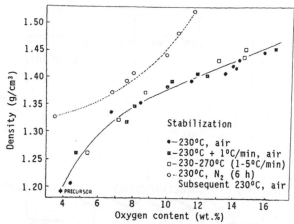

Figure 2.16 Correlation between density and oxygen content of stabilized PAN fibers. From Ref. 32. (By permission of the publishers, Butterworth–Heinemann Ltd.)

The stabilization reactions of cyclization, dehydrogenation, and oxidation are illustrated in Figure 2.17 [22]. Numerous gaseous by-products evolve during the pyrolysis, as shown in Figure 2.18 [32]. A possible mechanism for the evolution of HCN, NH₃, and H₂ is illustrated in Figure 2.19 [22]. The process has been modeled to describe the temperature and composition in the fibers during stabilization [33].

The duration of the stabilization must be sufficient for oxidation to take place throughout the entire cross section of the fibers; otherwise the unoxidized cores give rise to central holes in the carbon fibers. The oxidation is diffusion controlled (Figure 2.20). Stabilization in air usually takes several hours. For a PAN copolymer containing about 2% methacrylic acid, it takes only 25 min. [4]. An increase in the comonomer content reduces the time required for stabilization and improves the mechanical properties of the carbon fibers, but it does reduce their yield [22], as shown in Figure 2.21.

Prestabilization treatments are also used to reduce the stabilization time by decreasing the energy of activation of stabilization reactions. These treatments involve the impregnation of PAN precursor fibers with solutions of persulfate, cobalt salts, a combination of a salt of iron (II) and hydrogen peroxide, acids, guanidine carbonate, dibutylindimethoxide, and potassium permanganate [22].

An acidic medium (such as sulfur dioxide and hydrogen chloride) during stabilization causes an increase in the reaction rate and a greater degree of stabilization. This is due to a shift of the equilibrium of the stabilization reactions in the forward direction.

$$\text{acrylic precursor} \rightarrow \text{stabilized fiber} + NH_3 + HCN$$

$$NH_3 + HCl \rightarrow NH_4Cl$$

$$NH_3 + SO_2 + H_2O \rightarrow NH_4HSO_3$$

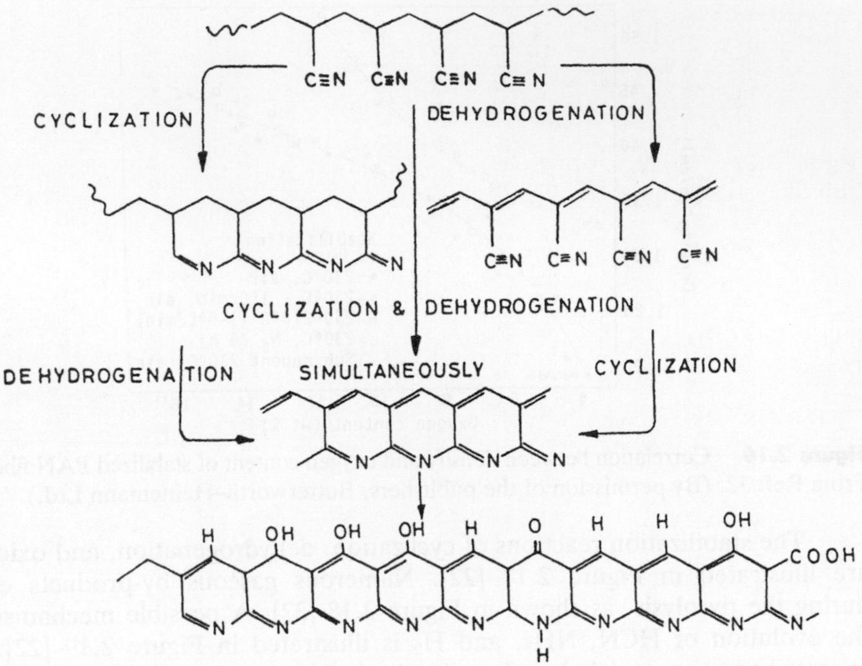

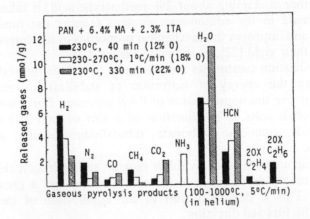

Figure 2.17 Sequence of reactions during thermooxidative stabilization of PAN. From Ref. 22. (By courtesy of Marcel Dekker Inc.)

Figure 2.18 Volatile by-products during pyrolysis of copolymeric PAN. From Ref. 32. (By permission of the publishers, Butterworth–Heinemann Ltd.)

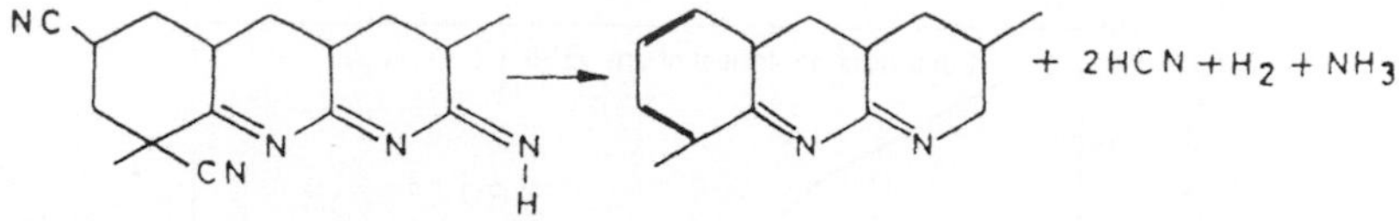

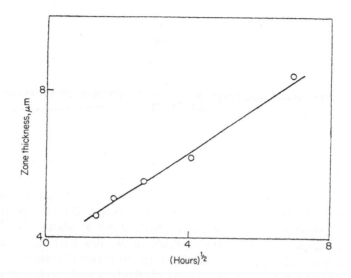

Figure 2.19 Mechanism of evolution of HCN, NH₃, and H₂ from the final stabilized structure of PAN. From Ref. 22. (By courtesy of Marcel Dekker Inc.)

Figure 2.20 Oxidized zone thickness as a function of $(time)^{1/2}$ for PAN fibers in air at 220°C. From Ref. 23. (Reprinted by courtesy of Elsevier Publishers B.V.)

The removal of ammonia as a salt shifts the equilibrium to the forward direction [22].

A fiber is taken as being properly stabilized when the oxygen content is 8–12%. An oxygen content in excess of 12% results in deterioration of the fiber quality, whereas an oxygen content below 8% results in a low carbon yield [4]. Due to the introduction of oxygenated groups and evolution of hydrogen cyanide, ammonia and other gases, the overall weight change during stabilization is small. However, at temperatures just above that of stabilization, significant weight loss can occur, especially if stabilization is not complete.

The density of the fiber increases from 1.17 g/cm³ for the original PAN fiber to about 1.40 g/cm³ for the stabilized fiber. However, the exact density depends on the precursor and the tension condition [22].

Stabilization is accompanied by a change in the color of the fiber from white, through shades of yellow and reddish brown, ultimately to shiny black.

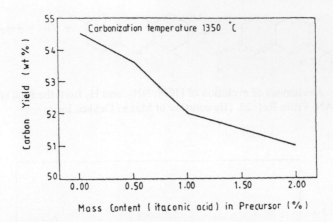

Figure 2.21 Influence of comonomer (itaconic acid) content on the carbon yield of PAN. From Ref. 22. (By courtesy of Marcel Dekker Inc.)

An adequately stabilized fiber resists chemical attack by mineral acids and bases, and does not burn when held inside a flame [27].

The shrinkage during stabilization consists of a physical contribution called entropy shrinkage, which is completed below 200°C, and a chemical contribution called reaction shrinkage, which starts at about 200°C. Entropy shrinkage is incipient contraction of PAN molecules that have been highly aligned during stretching prior to stabilization. Reaction shrinkage is due to the shortening of the PAN molecules during cyclization and oxygen-group formation. A higher copolymer content causes a larger chemical shrinkage. An increased heating rate (beyond 5°C/min.) enhances chemical shrinkage, while the entropy shrinkage remains unchanged. Thus, the optimum heating rate should be less than 5°C/min., i.e., 1–3°C/min. [4,22].

As the nitrile group gradually vanishes during stabilization, there is a transient ability for the polymer chains to slide past each other; this results in elongation after the initial shrinkage, as shown in Figure 2.22 for the straight-through path (stp), i.e., the case of a fiber under constant tension. Figure 2.22 also shows a constant-length path (clp), i.e., the case of a fiber kept at a constant length.

After stabilization, the fibers are carbonized or pyrolyzed by heating in an inert atmosphere (nitrogen) at 400–1 500°C. Tension is not required during carbonization as the all-carbon backbone of PAN remains largely intact after stabilization. In contrast, the rayon precursor has one oxygen atom in the backbone per monomer unit, so it undergoes considerable structural reorganization as the heteroatoms are lost during carbonization.

During carbonization about 50% by weight of the fiber is lost as gases such as water, ammonia, hydrogen cyanide, carbon monoxide, carbon dioxide, nitrogen, hydrogen, and possibly methane. The volume of gas evolved is 10^5

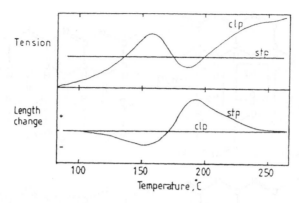

Figure 2.22 Typical autotension and length change profiles of PAN fiber versus temperature during oxidation for constant-length path (clp) and straight-through path (stp). From Ref. 29. (Reprinted by courtesy of Elsevier Science Publishers B.V.)

times the volume of the fibers [29]. Thus, an inert gas is used to dilute the toxic waste gas in the gas extract system, as well as to prevent ingress of atmospheric air. The treatment and disposal cost for the hydrogen cyanide by-product increases the production cost of PAN-based carbon fibers.

The rate of heating in the early stages of carbonization is low (less than 5°C/min. up to about 600°C) so that the release of volatiles is slow and does not cause pores or surface irregularities in the fiber. At 600–1 500°C, higher heating rates can be used because of the completion of the by-product evolution by 600°C, leaving only carbon (> 92 wt.%) and nitrogen (~ 6 wt.%). At 1 000–1 500°C, the residual nitrogen is progressively removed [4,29]. The overall residence time for carbonization is of the order of an hour, with residence at temperatures above 1 000°C of the order of minutes [29].

During carbonization, intermolecular cross-linking occurs through oxygen-containing groups (Figure 2.23) or through dehydrogenation (Figure 2.24), and the cyclized sections coalesce by cross-linking (Figure 2.25) to form a graphite-like structure in the lateral direction. The modulus starts to increase at 300°C [23]. Carbonization increases the fiber density from 1.45 to 1.70 g/cm^3 and decreases the fiber diameter from 10–15 μm to 6–9 μm [29].

Graphitization (optional) is carried out after carbonization by heating at 1 500–3 000°C in an inert atmosphere, which is nitrogen up to 2 000°C and argon above 2 000°C. Nitrogen cannot be used above 2 000°C because of the reaction between nitrogen and carbon to form cyanogen, which is toxic. A low cooling rate after the heating is preferred [22]. During graphitization, very little gas is evolved, but the crystallite size is increased and preferred orientation is improved, so the fiber becomes more graphitic. The residence time is just minutes for graphitization. The high temperatures make graphitization an expensive step, hence it is often skipped.

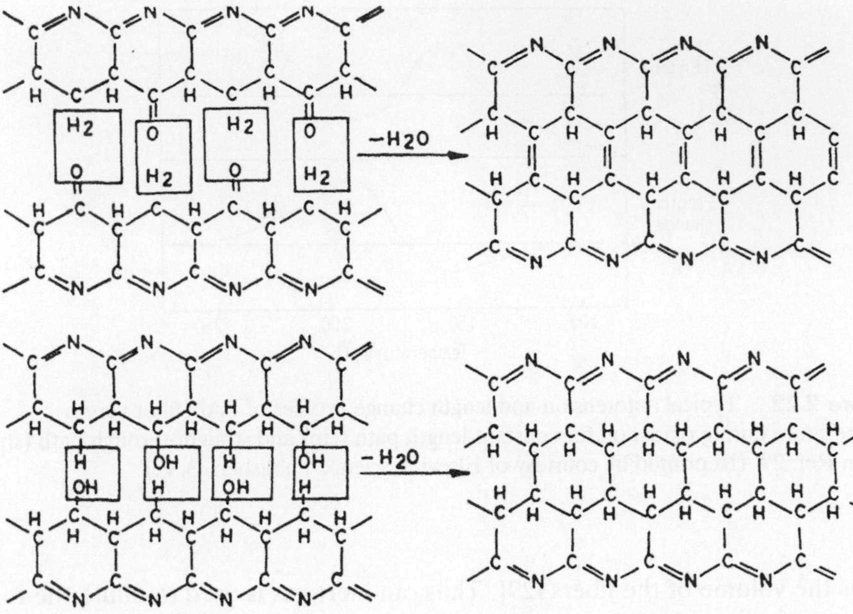

Figure 2.23 Intermolecular cross-linking of stabilized PAN fibers during carbonization through oxygen-containing groups. From Ref. 22. (By courtesy of Marcel Dekker Inc.)

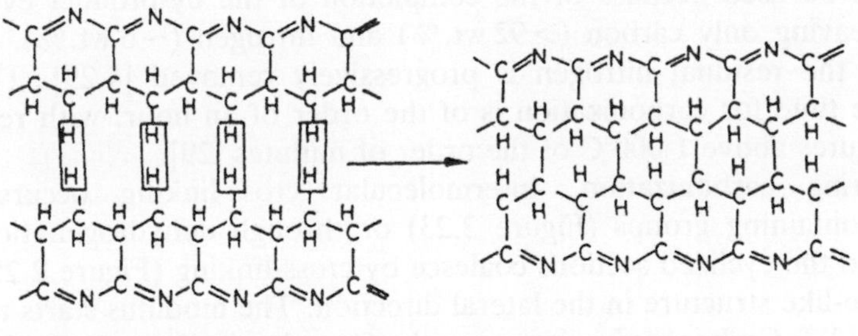

Figure 2.24 Intermolecular cross-linking of stabilized PAN fibers during carbonization through dehydrogenation. From Ref. 22. (By courtesy of Marcel Dekker Inc.)

Figure 2.26 shows the effect of the heat treatment temperature during carbonization and graphitization on the tensile strength and modulus of the resulting fiber. Zones A, B, and C in Figure 2.26 correspond approximately to Types III, II, and I in Figure 2.13, respectively. Type I carbon fibers have high modulus and low strength, hence low ductility. Type II fibers have lower modulus but higher strength, hence higher ductility. The processing cost is

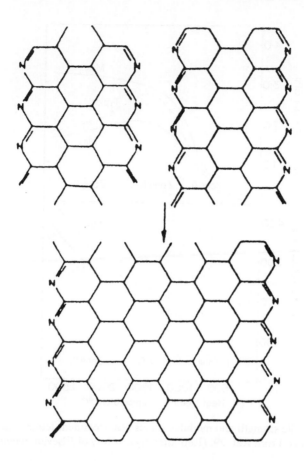

Figure 2.25 Cross-linking of the cyclized sequences in PAN fibers during carbonization. From Ref. 22. (By courtesy of Marcel Dekker Inc.)

much lower for Type II than for Type I fibers. For most structural applications, Type II fibers are used.

The long stabilization time required for PAN adds much to the cost of PAN-based carbon fibers. A polymer which does not require stabilization is poly(p-phenylene benzobisoxazole) (PBO). The use of this thermoplast for the production of carbon fibers is being investigated [34].

The high cost of PAN makes it attractive to use lower-cost polymers for making carbon fibers. An example of a low-cost polymer is polyethylene. Melt spun polyethylene fibers are cross-linked with chlorosulfonic acid for stabilization then carbonized at 900°C [35]. Carbon fibers with a tensile strength of 2.16 GPa, a modulus of 130 GPa, a high strain at break of 3%, and a diameter of 13 μm have been obtained from polyethylene [36].

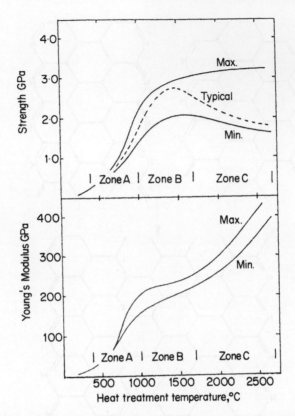

Figure 2.26 Tensile strength and modulus versus heat treatment temperature for stabilized PAN fibers. From Ref. 29. (Reprinted by courtesy of Elsevier Science Publishers B.V.)

The choice of the polymer can affect the microstructure of the resulting carbon fiber. By using a linear stiff-chain polymer, namely poly(p-phenylene benzobisthiazole) (PBZT), carbon fibers with well-defined fibrils along the fiber direction were obtained [37].

Carbon Fibers made from Carbonaceous Gases

Carbon filaments (to be distinguished from carbon fibers) grow catalytically when a carbonaceous gas is in contact with a small metal particle (the catalyst) at an elevated temperature [38]. During growth a carbon filament lengthens such that the filament diameter is equal to the diameter of the catalyst particle that produces it, as illustrated in Figure 2.27 [39]. While the filament lengthens by catalytic growth, noncatalytic chemical vapor deposition of carbon takes place from the carbonaceous gas on the sides of the filament, causing the filament to grow radially (thicken); it thus becomes a vapor grown carbon fiber (VGCF), also illustrated in Figure 2.27.

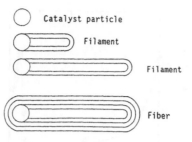

Figure 2.27 Diagram showing how a carbon filament is formed from a catalytic particle and how a carbon fiber is formed from a carbon filament. From Ref. 39. (Reprinted with permission from Pergamon Press plc.)

Iron is the most commonly used catalyst [40], though nickel [41], copper, palladium [42], and other metals and alloys can be used instead. The reactivity between the catalyst and its support (e.g., MgO, SiO_2, etc.) should be small, if any. For example, Pd supported on SiO_2 suffers from a reaction that forms Pd_2Si and this leads to suppression of the catalytic activity [42]. Iron particles can be obtained from solutions of iron salts or iron organometallics, listed in Table 2.7; the iron particle size is of the order of 10 nm. The addition of acetylacetonates of Fe, Co, and Mn to $Fe(C_5H_5)_2$ (ferrocene) reduces the size of the particles and consequently increases the filament growth rate and the filament yield [40]. In particular, the growth rate and yield are 40 μm/sec. and 70 wt.%, respectively, when 20 wt.% cobalt acetylacetonate is used along with 80 wt.% ferrocene [40]. Sulfurizing the iron, using thiophene or hydrogen sulfide as the sulfur source, helps penetration of catalyst into the fine pores of the activated carbon particles and allows growth of carbon filaments on to them [43]. This effect of sulfur is due to the molten state of the sulfurized iron [43]. A typical iron concentration is 5×10^{-6} g/cm^2 [44], as only a small percentage of catalyst particles gives filaments. When a catalyst particle becomes completely encased by the vapor deposited carbon, the catalyst is poisoned and the catalytic growth stops.

Table 2.7 Fibers and catalyst particles obtained from various precursors. From Ref. 44.

Precursor	Mean diameter of the fibers (μm)	Mean length (mm)	Mean diameter of the catalyst particles after germination (nm)
$Fe(NO_3)_3 \cdot 9H_2O$	3	1.11	11.2
$Fe(C_5H_5)_2$	5.1	0.55	35.7
$Fe_2(SO_4)_3$	3.2	0.78	25
$Fe(NO_3)_3 \cdot 9H_2O + KOH$	3.9	1.38	9

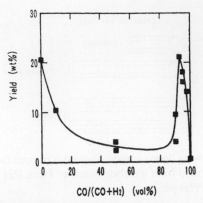

Figure 2.28 Effect of CO content in the CO_2/H_2 mixture on the carbon filament yield. From Ref. 45. (Reprinted with kind permission from Pergamon Press Ltd, Headington Hill Hall, Oxford OX3 0BW, UK.)

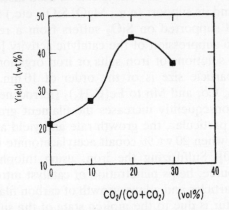

Figure 2.29 Effect of CO_2 content in the $CO/CO_2/H_2$ mixture on the carbon filament yield. From Ref. 45. (Reprinted with kind permission from Pergamon Press Ltd, Headington Hill Hall, Oxford OX3 0BW, UK.)

The carbonaceous gas can be acetylene, ethylene, methane, natural gas, benzene, etc. It is usually mixed with hydrogen during use. The addition of CO enhances the yield, if the $CO/(CO + H_2)$ volume fraction is 93–95%, as illustrated in Figure 2.28 [45]. This is because CO itself can reduce iron oxides, but the presence of both CO and H_2 in substantial quantities causes the reaction [45].

$$H_2 + CO \rightarrow H_2O + C$$

The water produced deactivates the catalyst iron [45]. The further addition of CO_2 to the CO/H_2 mixture enhances the yield. The highest yield, close to 45 wt.%, is obtained in a mixture of 77% CO, 19% CO_2 and 4% H_2—a composition similar to that of the Linz–Donawitz converter gas—as shown in Figure 2.29 [45] for the case of benzene as the carbon source. The

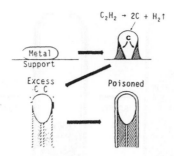

Figure 2.30 The most common mechanism of carbon filament formation. From Ref. 46. (Reprinted by permission of Kluwer Academic Publishers.)

effect of the CO_2 addition is tentatively attributed to either formation of fine catalyst particles or prevention of aggregation of the catalyst particles [45].

The mechanism of carbon filament formation is illustrated in Figure 2.30. The carbon-containing gas adsorbs and decomposes on the surface of the catalyst particle. The surface carbon then dissolves in the metal particle and diffuses through the particle from the hotter leading surface, on which hydrocarbon decomposition (an exothermic reaction) occurs, to the cooler rear face, at which carbon is precipitated from solution (an endothermic reaction). The carbon precipitate constitutes the growing filament. Thus, the catalyst particle is lifted up from its support as the filament grows and it remains at the top of the growing filament, leaving behind a hollow tube (20–500 Å in diameter) behind it in the center of the filament along the filament axis. Excess carbon, which accumulates at the exposed particle, is transported by surface diffusion around the peripheral surfaces of the particle to form the graphitic skin of the filament. Catalytic filament growth ceases when the leading face is encapsulated by a layer of carbon, which prevents further hydrocarbon decomposition [46].

The diffusion of carbon through a catalytic particle is the rate-limiting step of carbon filament growth. This is indicated by the observation that the rate of filament growth has an inverse square root dependence with the particle size [46]. Thus, too large a particle size will choke off the supply of carbon for the growing precipitate [47].

When the catalyst has been poisoned by the enveloping carbon, the catalytic activity can be restored by replacing the hydrocarbon by either hydrogen or oxygen and heating for a short time at 700°C. During hydrogasification (Figure 2.31), carbon diffuses in the direction opposite to that during filament growth and is subsequently converted to methane by reacting with adsorbed hydrogen at the front face of the particle. Thus, filament growth and hydrogasification are reversible processes [46].

When the catalyst particle interacts strongly with its support, it remains on the support as the carbon filament grows above it, as illustrated in Figure 2.32 for the case of Pt/Fe particles. In Figure 2.32, a Pt/Fe particle is modeled

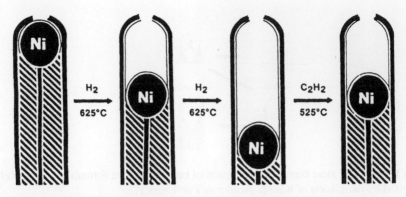

Figure 2.31 Reversible gasification/growth of carbon filaments from nickel particles. From Ref. 46. (Reprinted by permission of Kluwer Academic Publishers.)

as an Fe particle surrounded laterally by Pt, and acetylene (C_2H_2) is hypothesized to decompose on the Pt apron so that carbon diffuses to the Pt/Fe interface. After this point, the mechanism is the same as that in Figure 2.30. As the filament appears to be extruded from the catalyst particle, still attached to the support, this mode of growth is known as extruded filament growth; it is not as common as the growth depicted in Figure 2.30 [48].

The growth modes of Figures 2.30 and 2.32 both result in a carbon filament with a duplex structure; this consists of an internal and more reactive (less graphitic) core surrounded by a relatively oxidation-resistant (more graphitic) skin. As a result of this duplex structure, fibers in the form of hollow tubes can be obtained by gasifying the core [49].

Another way to obtain filaments in the form of hollow tubes involves using an additive (usually an oxide) on the catalyst particle to suppress the filament growth. Not only are the resulting filaments shorter, they tend to be hollow because less material diffuses through the catalyst particle to form the inner core of the filaments [46].

The growth modes depicted in Figures 2.30 and 2.32 result in whisker like filaments that grow in a single direction. But growth can occur in more than one direction to produce branched, bidirectional, and multidirectional filaments, as illustrated in Figure 2.33, where catalyst particles are indicated by solid figures. Branching does not necessarily require break up of a catalyst particle [46].

Carbon filaments are generally tubular in shape, but braided filaments and flat ribbons have been reported (Figure 2.34) [48]. In particular, coiled fibers with a thickness of 0.1–0.3 μm, a coil diameter of 2–8 μm, and a coil length of 0.1–5 mm have been obtained. They are formed by the intertwining of a pair of coils growing in the same direction simultaneously from a diamond-shaped Ni seed. They can be extended elastically up to about three times the original coil length [49]. On the other hand, for the growth of straight

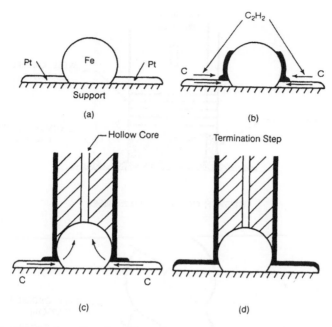

Figure 2.32 Mechanism of carbon filament formation by extruded filament growth. From Ref. 48. (Reprinted by permission of Kluwer Academic Publishers.)

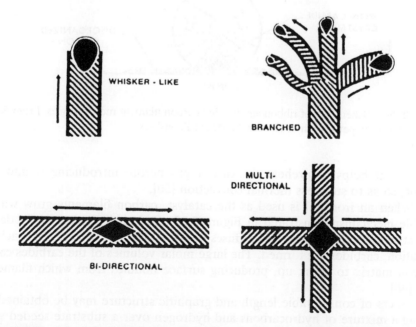

Figure 2.33 Schematic representation of different types of growth observed in carbon filaments. From Ref. 46. (Reprinted by permission of Kluwer Academic Publishers.)

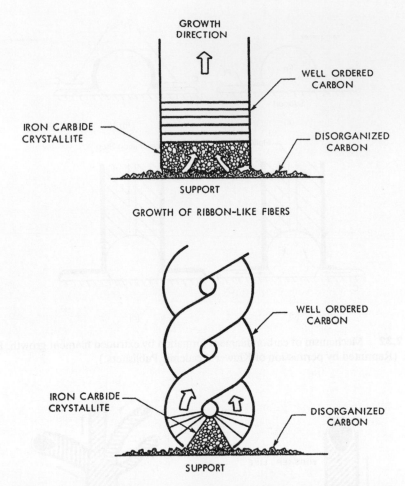

Figure 2.34 Schematics of ribbon and braided carbon filament morphologies. From Ref. 48. (Reprinted by permission of Kluwer Academic Publishers.)

filaments, it helps to preheat the carrier gas before introducing it into the reactor, so as to suppress thermal convection [50].

When an iron foil is used as the catalyst, carbon filaments grow with a possible mechanism illustrated in Figure 2.35. Carbon is preferentially deposited at surface dislocations and diffuses into the bulk of the iron. On reaching saturation, carbides are formed. The large molar volumes of the carbides cause the iron matrix to break up, producing surface nodules from which filaments grow [48].

Fibers of considerable length and graphitic structure may be obtained by flowing a mixture of hydrocarbons and hydrogen over a substrate seeded with catalytic particles and heated to near 1 000°C. A linearly increasing temperature sweep from 950 to 1 120°C at a rate of 7.1°C/min. is suitable [51]. A lower

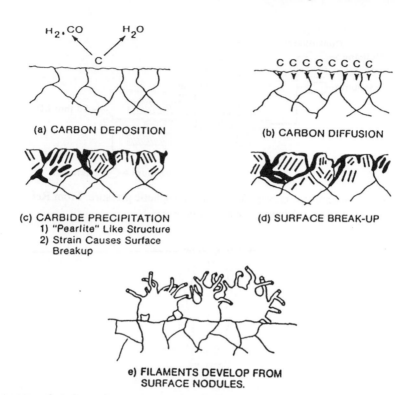

Figure 2.35 Initiation mechanism for development of carbon filaments on an α-iron foil. From Ref. 48. (Reprinted by permission of Kluwer Academic Publishers.)

hydrocarbon concentration is used for filament lengthening, whereas a higher hydrocarbon concentration is used for fiber thickening. For example, when methane is the source of carbon, the methane concentration in the gas stream is initially 5–15 vol.% for filament lengthening, and is later above 25 vol.% for fiber thickening [52]. Figure 2.36 illustrates the apparatus for growing VGCF at atmospheric pressure. Helium is used as an inert gas to minimize convection [53].

In an alternate configuration, catalyst particles or organometallics (such as ferrocene) are mixed with the hydrocarbon gas near the inlet of a reactor and sprayed into the reactor (Figure 2.37) so that filaments/fibers are grown in the gas stream as the stream progresses through the reactor. The filaments are of smaller diameter and length than the fibers grown on substrates, and their mechanical properties are inferior. However, this process has a high reactor productivity and is therefore commercially attractive [39,54]. A maximum initial growth rate of 1 500 μm/sec. has been reported [54].

The VGCF can be heat-treated in an inert atmosphere up to 3 000°C for graphitization. (Carbonization is not necessary.) As the catalyst particles are

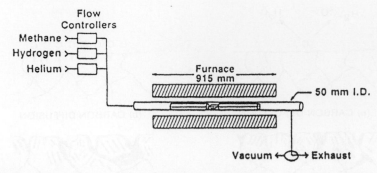

Figure 2.36 Apparatus for growing VGCF at atmospheric pressure. From Ref. 53. (Reprinted by permission of Kluwer Academic Publishers.)

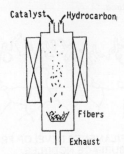

Figure 2.37 Apparatus for producing VGCF in large volume. From Ref. 39. (Reprinted with permission from Pergamon Press plc.)

completely encased in carbon at the conclusion of vapor deposition (thickening), they are not capable of catalytic effects during high-temperature heat treatments. Since the vapor deposited carbon has no metallic impurities and has a uniform cylindrical atomic layer microstructure, VGCF is highly graphitizable [53]. During graphitization, the growth of crystallites in the direction of the fiber axis lags behind the growth of crystallites perpendicular to the fiber axis [55].

Instead of a carbonaceous gas, the carbon source may be a plasma generated from an electrical arc produced from a carbon anode. As long as there is a source of carbon, the catalytic growth of carbon filaments is possible [56].

Carbon fibers can be formed from carbonaceous gases without using a catalyst. For example, carbon fibers in the form of tubules (their ends covered by carbon layers instead of the catalyst particle at the tip of a catalytically grown filament) can be grown from carbonaceous gases along with the formation of buckminsterfullerenes. The presence of carbon pentagons, in addition to carbon hexagons, at the end of a tubule allows it to be covered by carbon layers. Carbon tubules are typically smaller in diameter than catalytically grown carbon filaments. The number of carbon layers is around eight in a

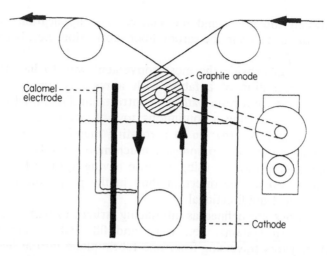

Figure 2.38 Schematic of a typical arrangement for continuous anodic oxidation of carbon fibers. From Ref. 24. (By permission of Elsevier Science Publishers B.V.)

carbon tubule [57]. As another example, carbon fibers are grown from a carbonaceous gas that causes the deposition of carbon at the tips of growing fibers electrostatically separated from one another [58]. Carbon fibers can also grow at the ends of corona wires during negative point-to-plane corona discharges in hydrocarbon atmospheres [59].

Surface Treatment of Carbon Fibers

Surface treatment of carbon fibers is performed for improving the adhesion between the fibers and the matrix when the fibers are used in a composite. The preferred type of treatment depends on the matrix material, which can be a polymer (a thermoset or a thermoplast), a metal, a carbon, or a ceramic. The details for each matrix are described in the chapter of that title. In general, there are three main methods of surface treatment, namely wet oxidation (e.g., HNO_3, 110°C, 10 min. to 150 h.), dry oxidation (e.g., air, O_2, 500–800°C, 30 sec. to 2 h.) and anodic oxidation (e.g., H_2SO_4, K_2SO_4, NaOH, 1–10 min.) [24]. In anodic oxidation, the anodic potential must be below that for the active oxidation reaction, which causes serious damage to the fibers (the anode). Figure 2.38 [24] is a schematic of a typical setup for continuous anodic oxidation of carbon fibers.

Carbon Fiber Fabrics, Preforms, and Staple Yarns

The brittleness of carbon fibers makes it necessary to coat them with a sizing before weaving. The choice of the sizing material depends on the matrix in which the fibers will be used, as the sizing also serves to enhance the bonding between the fibers and the matrix. The application of a sizing is carried out as the last step in the carbon fiber fabrication, i.e., before the fibers are wound on a spool.

Woven fabrics provide the most convenient way to handle continuous fibers. The impregnation of a fabric with a resin is simply performed by immersing the fabric in a resin bath. In contrast, unidirectional fiber prepregs are prepared by immersing a continuous tow of fibers in a resin bath and subsequently winding the tow on to a rotating mandrel. The equipment for preparing unidirectional fiber prepregs is much more complicated than that for preparing fabric prepregs. By using a fabric having a much larger number of yarns (twisted tows) in one direction than in the perpendicular direction, a fabric can be almost unidirectional.

Woven fabrics are defined as interlacing structures with at least two sets of orthogonal yarn systems, i.e., warp and fill. Fabrics are usually two-dimensional. If yarns traversing from one fabric plane into at least one other fabric plane exist (by means of structural stitching [60]), the fabric is three-dimensional and is known as a fabric preform [61].

Braided fabrics are defined as interlacing structures that consist of at least two sets of yarns, not necessarily orthogonal. For simple two-dimensional biaxial braids, the yarn orientations are generally $\pm\phi$, where ϕ can vary from approximately 15 to 89°. In addition to the angular oriented yarns, yarns oriented along the axial or machine direction of the braid (i.e., $\phi = 0$) can be added. These braids are referred to as two-dimensional triaxial braids. If more than one layer of fabric is braided at a time and the layers are interconnected so that yarns from one layer traverse to no less than at least one contiguous layer, the braid is referred to as a three-dimensional braid [61]. Braiding is particularly suitable for fabricating composite tubings from braided seamless sleeves.

The attraction of three-dimensional weaving is the increase in the interlaminar shear strength of the resulting composite, although this increase is accompanied by a decrease in the in-plane tensile modulus and strength.

Fabric preforms provide a way to fabricate composite materials of near net shapes. Furthermore, they can be used to build structures with a reduced amount of external fasteners. Fastenless preform shapes include I's, T's, and J's. A reduced amount of fasteners is attractive because the interface between a fastener and a composite is the "weak link" in a structure [61].

The fabrics and preforms mentioned above are anisotropic. Three-dimensionally isotropic preforms can be prepared by the natural intertwining (self-weaving) of carbon filaments as they grow by catalytic decomposition of a carbonaceous gas. By using a shaped mold for the carbon filament growth, preforms of near net shapes can be fabricated [62].

Three-dimensionally or two-dimensionally isotropic preforms can be prepared from short carbon fibers. A binder is used to make the short carbon fiber preform. The binder can be carbon (from a pitch or a resin), silica, phosphate, epoxy, etc. The fabrication of a preform involves (1) filtering or pressing a slurry consisting of short carbon fibers, a binder, and a vehicle, (2) drying, and (3) heat treatment (in some cases). In the case of a pitch binder, the pitch can be dissolved in toluene or, more safely, methylene chloride, and combined with carbon fibers to form a slurry. Water is the usual vehicle for silica or phosphate binders. Preforms prepared by dry pressing of short carbon fibers suffer from nonuniformity in the fiber distribution. The slurry method is particularly important for preforms with carbon fibers in amounts less than 25 vol.%.

A staple yarn consists of a series of short fibers that are placed to form a continuous overlapping assembly; the assembly is bound together by twist, which prevents fiber slippage by increasing the interfiber friction. The mean staple length of the fibers is about 100 mm (about 4 in.) in the Heltra process, which produces staple yarns from continuous carbon fiber tows. This length is well above the critical length associated with composite materials. In spite of the discontinuity and twist in the fibers, there is a high translation of tensile properties from the continuous fiber tow to the staple yarn; up to 97% of the tensile modulus and 84% of the tensile strength are retained. Due to the hairiness of a yarn, the use of a yarn in a composite significantly increases the interlaminar shear strength over the continuous fiber counterpart [63].

References

1. I.C. Lewis, *J. Chim. Physique* **81**, 751–758 (1984).
2. D.D. Edie, in *Carbon Fibers Filaments and Composites*, edited by J.L. Figueiredo, C.A. Bernardo, R.T.K. Baker, and K.J. Huttinger, Kluwer Academic, Dordrecht, 1990, pp. 43–72.
3. Kensuke Okuda, *Trans. Mater. Res. Soc. Jpn.* **1**, 119–139 (1990).
4. J.-B. Donnet and R. C. Bansal, *International Fiber Science and Technology, 10 (Carbon Fibers)*, 2d ed., Marcel Dekker, New York, 1990, Ch. 1.
5. Union Carbide, U.S. Patent 4,005,183 (1977).
6. S. Otani and A. Oya, in *Composites '86: Recent Advances in Japan and the United States, Proc. Japan–U.S. CCM-III*, edited by K. Kawata, S. Umekawa, and A. Kobayashi, Jpn. Soc. Compos. Mater., Tokyo, 1986, pp. 1–10.
7. Exxon Research & Engineering, U.S. Patent 4,277,325 (1980/81).
8. I. Seo, S. Takahashi, and T. Ohono, U.S. Patent 4,986,893 (1991).
9. B. Rhee, D.H. Chung, S.J. In, and D.D. Edie, *Carbon* **29**(3), 343–350 (1991).
10. T.-W. Fu and M. Katz, U.S. Patent 4,999,099 (1991).
11. L.S. Singer, U.S. Patent 4,005,183.
12. Y. Takai, M. Takabatake, H. Nakajima, K. Takamo, and M. Watanabe, U.S. Patent 4,913,889 (1990).
13. D.D. Edie, N.K. Fox, B.C. Barnett, and C.C. Fain, *Carbon* **24**(4), 477–482 (1986).
14. M. Hein, *Erdoel Kohle, Erdgas, Petrochem.* **43**(9), 354–358 (1990).
15. J. G. Lavin, *Carbon* **30**(3), 351–357 (1992).
16. I. Mochida, H. Tosnima, Y. Korai, and T. Hino, *J. Mater. Sci.* **24**(2), 389–94 (1989).
17. T. Matsumoto and I. Mochida, *Carbon* **31**(1), 143–147 (1993).

18. I. Mochida, S.-M. Zeng, Y. Korai, T. Hino, and H. Toshima, *Carbon* **29**(1), 23–29 (1991).
19. S.-M. Zeng, Y. Korai, I. Mochida, T. Hino, and H. Toshima, *Bull. Chem. Soc. Jpn.* **63**(7), 2083–2088 (1990).
20. S. Koga, T. Okajima, S. Yamaguchi, and E. Kakikura, European Patent Application EP 133457 A1 (1985).
21. E.M. Kohn, U.S. Patent 3,972,968 (1976).
22. A.K. Gupta, D.K. Paliwal, and Pushpa Bajaj, *J. Macromol. Sci., Rev. Macromol. Chem. Phys.* **C31**(1), 1–89 (1991).
23. W. Watt, in *Strong Fibers*, edited by W. Watt and B.V. Perov, North-Holland, Amsterdam, 1985, pp. 327–387.
24. E. Fitzer and M. Heine, in *Composite Materials Series, Vol. 2, Fibre Reinforced Composite Materials*, edited by A.R. Bunsell, Elsevier, Amsterdam, 1988, pp. 73–148.
25. G.P. Daumit, Y.S. Ko, C.R. Slater, J.G. Venner, C.C. Young, and M.M. Zurick, U.S. Patent 4,933,128 (1990).
26. G.P. Daumit, Y.S. Ko, C.R. Slater, J.G. Venner, D.W. Wilson, C.C. Young, and H. Zabaleta, in *Proc Int. SAMPE Tech. Conf., 20, Materials Processes: Intercept Point*, 1988, pp. 414–422.
27 S. Damodaran, P. Desai, and A.S. Abhiraman, *J. Text. Inst.* **81**(4), 384–420 (1990).
28. G.P. Daumit, J.D. Rector, J.G. Venner, D.W. Wilson, and C.C. Young, in *Proc. Int. SAMPE Symp. and Exhib., 35, Advanced Materials: Challenge Next Decade*, edited by G. Janicki, V. Bailey, and H. Schjelderup, 1990, pp. 1–12.
29. D.J. Thorne, in *Strong Fibers*, edited by W. Watt and B.V. Perov, North-Holland, Amsterdam, 1985, pp. 475–493.
30. P. Rajalingam and G. Radhakrishnan, *J. Macromol. Sci., Rev. Macromol. Chem. Phys.*, **C31**(2–3), 301–310 (1991).
31. R.M. Kimmel, J.P. Riggs, R.W. Swander, and W. Whitney, U.S. Patent 3,925,524 (1975).
32. E. Fitzer and W. Frohs, *Chem. Eng. Technol.* **13**(1), 41–49 (1990).
33. M.G. Dunham and D.D. Edie, *Carbon* **30**(3), 435–450 (1992).
34. D.D. Edie, private communication (1992).
35. A.R. Postema, H. De Groot, and A.J. Pennings, *J. Mater. Sci.* **25**, 4216–4222 (1990).
36. J.P. Pennings, R. Lagcher, and A.J. Pennings, *Polym. Bull. (Berlin)* **25**(3), 405–412 (1991).
37. H. Jiang, P. Desai, S. Kumar, and A.S. Abhiraman, *Carbon* **29**(4–5), 635–644 (1991).
38. H.G. Tennent, J.J. Barber, and R. Hoch, U.S. Patent 5,165,909 (1992).
39. G.G. Tibbetts, *Carbon* **27**(5), 745–747 (1989).
40. M. Ishioka, T. Okada, K. Matsubara and M. Endo, *Carbon* **30**(6), 865–868 (1992).
41. D.J. Smith, M.R. McCartney, E. Tracz, and T. Borowiecki, *Ultramicroscopy* **34**(1–2), 54–59 (1990).
42. L. Kepinski, *React. Kinet. Catal. Lett.* **38**(2), 363–367 (1989).
43. T. Kato, K. Haruta, K. Kusakabe, and S. Morooka, *Carbon* **30**(7), 989–994 (1992).
44. P. Gadelle, in *Carbon Fibers Filaments and Composites*, edited by J.L. Figueiredo, C.A. Bernardo, R.T.K. Baker, and K.J. Huttinger, Kluwer Academic, Dordrecht, 1990, pp. 95–117.
45. M. Ishioka, T. Okada, K. Matsubara, and M. Endo, *Carbon* **30**(6), 859–863 (1992).

46. R.T.K. Baker, in *Carbon Fibers Filaments and Composites*, edited by J.L. Figueiredo, C.A. Bernardo, R.T.K. Baker, and K.J. Huttinger, Kluwer Academic, Dordrecht, 1990, pp. 405–439.
47. G.G. Tibbetts, in *Carbon Fibers Filaments and Composites*, edited by J.L. Figueiredo, C.A. Bernardo, R.T.K. Baker, and K.J. Huttinger, Kluwer Academic, Dordrecht, 1990, pp. 525–540.
48. A. Sacco, Jr., in *Carbon Fibers Filaments and Composites*, edited by J.L. Figueiredo, C.A. Bernardo, R.T.K. Baker, and K.J. Huttinger, Kluwer Academic, Dordrecht, 1990, pp. 459–505.
49. S. Motojima, M. Kawaguchi, K. Nozaki and H. Iwanaga, *Carbon* **29**(3), 379–385 (1991); *Appl. Phys. Lett.* **56**(4), 321–323 (1990).
50. M. Ishioka, T. Okada, and K. Matsubara, *Carbon* **31**(1), 123–127 (1993).
51. G.G. Tibbetts, *Carbon* **30**(3), 399–406 (1992).
52. G.G. Tibbetts, U.S. Patent 4,565,684 (1986).
53. G.G. Tibbetts, in *Carbon Fibers Filaments and Composites*, edited by J.L. Figueiredo, C.A. Bernardo, R.T.K. Baker, and K.J. Huttinger, Kluwer Academic, Dordrecht, 1990, pp. 73–94.
54. T. Masuda, S.R. Mukai, and K. Hashimoto, *Carbon* **30**(1), 124–126 (1992).
55. K.K. Brito, D.P. Anderson, and B.P. Rice, in *Proc. Int. SAMPE Symp. and Exhib., 34, Tomorrow's Materials: Today*, edited by G.A. Zakrzewski, D. Mazenko, S.T. Peters, and C.D. Dean, 1989, pp. 190–201.
56. J.D. Fitz Gerald, G.H. Taylor, L.F. Brunckhorst, L.S.K. Pang, and M.A. Wilson, *Carbon* **31**(1), 240–244 (1993).
57. D. Cox, private communication.
58. J.P. Glass, U.S. Patent 3,915,663 (1975); U.S. Patent 3,536,519 (1970).
59. J.R. Brock and P. Lim, *Appl. Phys. Lett.* **58**(12), 1259–1261 (1991).
60. A. Morales, in *Proc. 22nd Int. SAMPE Tech. Conf.*, 1990, pp. 1217–1230.
61. D. Brookstein, *J. Appl. Polym. Sci., Appl. Polym. Symp.* **47**, 487–500 (1991).
62. H. Witzke and B.H. Kear, U.S. Patent 4,900,483 (1990).
63. R.J. Coldicott and T. Longdon, in *Proc. Int. SAMPE Symp. and Exhib., 34, Tomorrow's Materials: Today*, edited by G.A. Zakrzewski, D. Mazenko, S.T. Peters and C.D. Dean, 1989, pp. 202–210.

CHAPTER 3

Structure of Carbon Fibers

The properties of carbon fibers strongly depend on the structure. The properties include tensile modulus, tensile strength, electrical resistivity, and thermal conductivity. The structural aspects that are particularly important are (1) the degree of crystallinity, (2) the interlayer spacing (d_{002}), (3) the crystallite sizes or, more accurately, the coherent lengths perpendicular (L_c) and parallel (L_a) to the carbon layers, (4) the texture (preferred orientation of the carbon layers) parallel and perpendicular to the fiber axis, (5) the transverse and longitudinal radii of curvature (r_t and r_l) of the carbon layers, (6) the domain structure, and (7) the volume fraction, shape and orientation of microvoids. A high degree of crystallinity, a low interlayer spacing, large crystallite sizes, a strong texture parallel to the fiber axis, and a low density of in-plane defects (disclinations) generally result in a high tensile modulus, a low electrical resistivity, and a high thermal conductivity. A weak texture perpendicular to the fiber axis, small values of r_t and r_l, a large amount of defects and distortions within a layer, a large value of L_c, and a low volume fraction of microvoids contribute to a high tensile strength. However, a large L_c value may be accompanied by reduced lateral bonding between the stacks of carbon layers, thereby degrading the strength. The structure is affected by the processing of the fibers, particularly the heat treatment temperature and the ease of graphitization of the carbon fiber precursor. PAN-based carbon fibers, even after heat treatment beyond 2 000°C, remain turbostratic (i.e., no graphitic ABAB stacking of the carbon layers) [1].

The interlayer spacing decreases while L_c and L_a increase with the heat treatment temperature. Figure 3.1 [2] shows that L_c increases with the heat treatment temperature, such that its value is higher for pitch-based carbon fibers than PAN-based carbon fibers that have been heat-treated at the same temperature. For PAN-based fibers, L_c increases sharply above 2 200°C, whereas L_c increases smoothly with increasing temperature for pitch-based fibers.

Figure 3.2 [2] shows L_a parallel to the fiber axis (i.e., $L_{a\parallel}$) and L_a perpendicular to the fiber axis (i.e., $L_{a\perp}$), both plotted against L_c. Both $L_{a\parallel}$ and

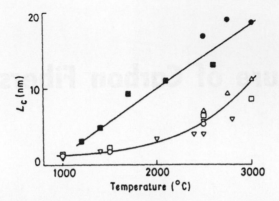

Figure 3.1 Plots of L_c against the heat treatment temperature for pitch-based carbon fibers (solid symbols) and PAN-based carbon fibers (open symbols). From Ref. 2.

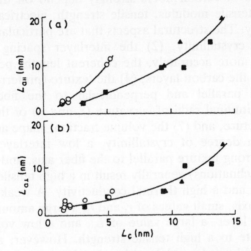

Figure 3.2 Plots of (a) $L_{a\parallel}$ and (b) $L_{a\perp}$ against L_c for pitch-based carbon fibers (solid symbols) and PAN-based carbon fibers (open symbols). From Ref. 2.

$L_{a\perp}$ increase with increasing L_c, but $L_{a\parallel}$ is larger than $L_{a\perp}$ for the same value of L_c. The value of $L_{a\perp}$ is slightly less than that of L_c, whereas that of $L_{a\parallel}$ can be larger than that of L_c.

Figure 3.3 [2] shows the volume fractions of crystallites (v_c), of unorganized or noncrystalline carbons (v_a), and of microvoids (v_p) versus L_c. The volume fraction of crystallites (v_c) describes the degree of crystallinity; it increases with increasing L_c. An increase in v_c from 50 to 98% is accompanied by a decrease in v_a from 50 to 0% and an increase in v_p from 0 to 2% for pitch-based carbon fibers. In spite of the increase in v_p with increasing L_c, the fiber density increases with L_c for pitch-based carbon fibers (Figure 3.4 [2]) due

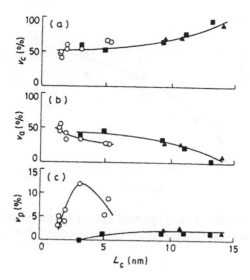

Figure 3.3 Plots of (a) v_c, (b) v_a, and (c) v_p against L_c for pitch-based carbon fibers (solid symbols) and PAN-based carbon fibers (open symbols). From Ref. 2.

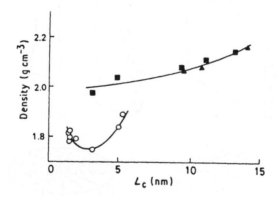

Figure 3.4 Fiber density plotted against L_c for pitch-based carbon fibers (solid symbols) and PAN-based carbon fibers (open symbols). From Ref. 2.

to the fact that crystalline carbon has a higher density than unorganized carbon. Figure 3.4 also shows that pitch-based carbon fibers have a higher density than PAN-based carbon fibers; this is due to the higher value of v_p in PAN-based carbon fibers (Figure 3.3).

The degree of orientation (f) of the carbon layers parallel to the fiber axis increases with $L_{a\parallel}$, as shown in Figure 3.5 [2]. Even for the highest $L_{a\parallel}$ of 20 nm, f is only 0.96. The parameter f is a description of the texture along the fiber axis.

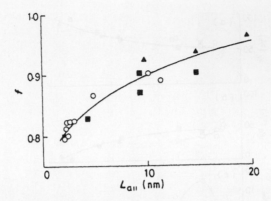

Figure 3.5 Degree of orientation (*f*) plotted against $L_{a\parallel}$ for pitch-based carbon fibers (solid symbols) and PAN-based carbon fibers (open symbols). Ref. 2.

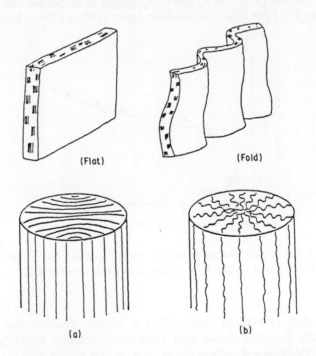

Figure 3.6 Texture models of mesophase pitch-based carbon fibers: (a) oriented core structure of Thornel and (b) folded layer structure of Carbonic. From Ref. 3.

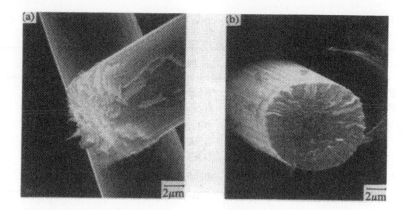

Figure 3.7 SEM photographs of Thornel P-100 carbon fibers: (a) as received, side view and (b) partially oxidized (17% weight loss), tip view.

The texture perpendicular to the fiber axis varies; it depends on the fiber processing conditions. In the case of pitch-based carbon fibers, the spinning conditions control the turbulence in the pitch during spinning and the turbulence in turn affects the texture. Figure 3.6 [3] shows the texture perpendicular to the fiber axis (i.e., in the cross-sectional plane of the fiber) of mesophase pitch-based carbon fibers from two sources, namely Amoco Performance Products (Thornel P-100, P-120) and Kashima Oil Co. (Carbonic HM50, HM60, HM80). Thornel has an oriented core structure, with relatively flat carbon layers and extremely strong texture along the fiber axis, as shown by the SEM photographs in Figure 3.7. (In Figure 3.7b oxidation has partially removed the skin on the fiber, thus revealing the texture more clearly.) Thornel has a well-developed three-dimensional graphitic structure. Carbonic has a folded layer structure, which results in a turbostratic (not graphitic) layer structure, even after heat treatment at $2\,850°C$. Moreover, the texture of Carbonic along the fiber axis is not as strong as that of Thornel. As a result, d_{002} is larger for Carbonic than Thornel, while L_c is smaller for Carbonic than Thornel. For comparison, Table 3.1 shows d_{002} and L_c of Torayca M46, which is a high-modulus PAN-based carbon fiber made by Toray. The low value of L_c and the high value of d_{002} for Torayca is consistent with the fact that PAN is not as graphitizable as pitch. Table 3.2 shows the corresponding tensile properties of Thornel, Carbonic, and Torayca fibers. Thornel fibers have low strength, high modulus, and low elongation (i.e., ductility); Carbonic fibers have high strength, low modulus (except for HM80), and high elongation; Torayca fibers have low strength and low modulus. The low strength of Thornel is attributed to the flat layer structure (Figure 3.6a), which facilitates crack propagation (Figure 3.8a). The high strength of Carbonic is attributed to the folded layer structure (Figure 3.6b), which increases the resistance to crack propagation (Figure 3.8b). The higher modulus of Thornel compared to Carbonic is due to the stronger texture of Thornel along the fiber axis.

Table 3.1 Structural parameters determined by X-ray diffraction on various carbon fibers. From Ref. 3.

Sample name	d_{002} (nm)	L_C (nm)
Thornel P-100	0.3392	24
P-120	0.3378	28
Carbonic HM50	0.3423	13
HM60	0.3416	15
HM80	0.3399	18
Tarayca M46	0.3434	6.2

Table 3.2 Mechanical properties of various carbon fibers. From Ref. 3.

Sample name	Tensile strength (GPa)	Tensile modulus (GPa)	Elongation (%)
Thornel P-100	2.2	690	0.3
P-120	2.4	830	0.3
Carbonic HM50	2.8	490	0.6
HM60	3.0	590	0.5
HM80	3.5	790	0.4
Torayca M46	2.4	450	0.5

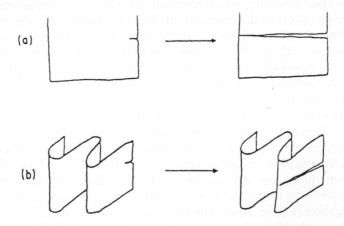

Figure 3.8 Fracture models of mesophase pitch-based carbon fibers with (a) oriented core structure and (b) folded layer structure. From Ref. 3.

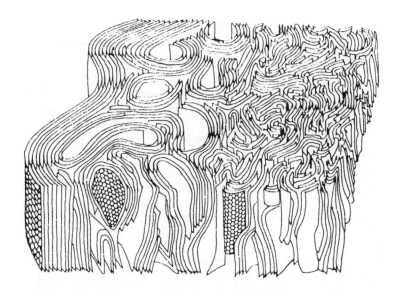

Figure 3.9   Texture model of Type I PAN-based carbon fibers showing skin–core heterogeneity. From Ref. 4.

The cross-sectional texture of a carbon fiber can be different between the core and the skin of the fiber, whether the cross-sectional shape of the fiber is round, dog-bone, or others. For PAN-based carbon fibers, the skin tends to have the carbon layers lined up parallel to the perimeter of the fiber, whereas the core tends to have the carbon layers exhibiting a random cross-sectional texture (also called a turbostratic texture), as illustrated in Figure 3.9 [4]. This duplex structure (skin–core heterogeneity) is characteristic of Type I PAN-based carbon fibers (e.g., those with a tensile strength of 1.9 GPa and a tensile modulus of 517 GPa [4]). The formation of a skin is probably the result of layer-plane ordering, which occurs as the heat treatment temperature is increased; the fiber surface presents a constraint on the number of possible orientations of surface crystallites [5]. The development of the skin structure is illustrated schematically in Figure 3.10 [6] for heat treatments at 1 000, 1 500, and 2 500°C. The surface layers become more ordered and continuous as heating proceeds, until the skin becomes continuous after heating at 2 500°C. After heating at 2 500°C, the skin is 1.5 μm thick and the core is 3 μm in diameter [5]. The skin–core heterogeneity can also be due to the higher temperature at the skin than the core during carbonization, and the larger stretching force experienced by the skin than the core [7].

For Type II PAN-based fibers (e.g., a tensile strength of 3.7 GPa and tensile modulus of 240 GPa), there is no skin and the crystallites are smaller than those of Type I [4]. The smaller crystallites of Type II (300 × 200 × 1 000 Å at the surface, as shown by scanning tunneling micros-

Fiber axis

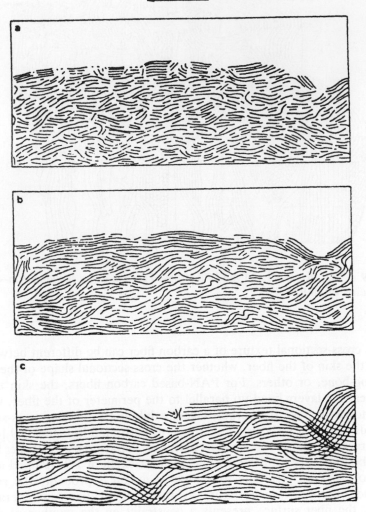

Figure 3.10 Schematic representation of the development of a skin from PAN-based carbon fibers heat-treated at (a) 1 000°C, (b) 1 500°C, and (c) 2 500°C. From Ref. 6.

copy [8]) causes the strength to be higher for Type II than Type I. In Type II, a high proportion of the carbon layers are not parallel to the fiber's cylindrical surface; instead they protrude [8]. This edge exposure enhances the bonding of the fiber to the matrix when the fiber is used in a composite. The stronger texture along the fiber axis gives Type I a higher modulus than Type II. The random cross-sectional texture of PAN-based fibers contrasts with the more sheetlike structure (whether oriented core structure or radial structure) of pitch-based carbon fibers. This random structure causes Type II PAN-based carbon fibers to exhibit higher strength than pitch-based carbon fibers [4].

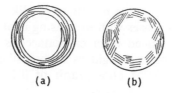

Figure 3.11 Schematics of (a) smooth laminar and (b) rough laminar skin structures. From Ref. 9.

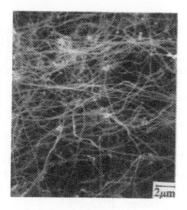

Figure 3.12 SEM photograph of carbon filaments of diameter 0.1–0.2 μm.

There are in general two types of skin structures: the smooth laminar structure has the carbon layers oriented as a smooth, cylindrical sheath parallel to the fiber surface (Figure 3.11a); the rough laminar structure has the carbon layers in the form of small crystallites, such that the layers are flat and parallel within each crystallite and the crystallites are roughly parallel to the fiber surface (Figure 3.11b) [9].

A mesophase pitch-based carbon fiber contains needlelike domains (microfibrils) up to 0.5 μm across, such that the needles are parallel or almost parallel to the fiber axis [10]. The domains are of two main types, namely dense domains (resulting from the mesophase portion of the pitch [10] and having an oriented texture [11]) and microporous domains (resulting from the isotropic pitch portion during spinning, such that this portion later yields some mesophase and volatiles [10], and having a random texture [11]). Graphite resides in the dense domains, such that the thickness of a graphite crystallite is much less than that of a domain [10]. The domains reside in a matrix which is turbostratic in texture [9]. The two types of domains form a zigzag nanostructure [12]. This domain structure is also known as pitch structure [9].

The pores in carbon fibers are mostly needlelike and elongated along the fiber axis [7]. Their size increases with increasing heat treatment temperature.

Carbon filaments produced from carbonaceous gases tend to be much smaller in diameter and not as straight as carbon fibers (pitch based or PAN based). Figure 3.12 is an SEM photograph of $0.1-0.2\,\mu m$ diameter carbon filaments fabricated by General Motors.

The sizing on carbon fibers is often an epoxy resin. Its imaging may be achieved by staining, which involves producing RuO_4 from $RuCl_3$ and sodium hypochlorite [13].

References

1. A. Duerbergue and A. Oberlin, *Carbon* **30**(7), 981–987 (1992).
2. A. Takaku and M. Shioya, *J. Mater. Sci.* **25**, 4873 (1990).
3. M. Endo, *J. Mater. Sci.* **23**, 598 (1988).
4. D.J. Johnson, *Chemistry and Industry* **18**, 692–698 (1982).
5. W. Johnson, in *Strong Fibres*, edited by W. Watt and B.V. Perov, North-Holland, Amsterdam, 1985, pp. 389–443.
6. S.C. Bennet, *Strength Structure Relationships in Carbon Fibers*, Ph.D. thesis, University of Leeds, 1976.
7. A.K. Gupta, D.K. Paliwal, and P. Bajaj, *J. Macromol. Sci., Rev. Macromol. Chem. Phys.* **C31**(1), 1–89 (1991).
8. P. Marshall and J. Price, *Composites* **22**(5), 388–393 (1991).
9. D.K. Brown and W.M. Phillips, in *Proc. Int. SAMPE Symp. and Exhib.*, 35, *Advanced Materials: Challenge Next Decade*, edited by G. Janicki, V. Bailey, and H. Schjelderup, 1990, pp. 2052–2063.
10. J.D. Fitz Gerald, G.M. Pennock, and G.H. Taylor, *Carbon* **29**(2), 139–164 (1991).
11. M. Inagaki, N. Iwashita, Y. Hishiyama, Y. Kaburagi, A. Yoshida, A. Oberlin, K. Lafdi, S. Bonnamy, and Y. Yamada, *Tanso* **147**, 57 (1991).
12. K. Lafdi, S. Bonnamy, and A. Oberlin, *Carbon* **31**(1), 29–34 (1993).
13. P. Le Coustumer, K. Lafdi, and A. Oberlin, *Carbon* **30**(7), 1127–1129 (1992).

CHAPTER 4

Properties of Carbon Fibers

Introduction

The properties of carbon fibers vary widely depending on the structure (Chapter 3) of the fibers. In general, attractive properties of carbon fibers include the following:

- low density
- high tensile modulus and strength
- low thermal expansion coefficient
- thermal stability in the absence of oxygen to over 3 000°C
- excellent creep resistance
- chemical stability, particularly in strong acids
- biocompatibility
- high thermal conductivity
- low electrical resistivity
- availability in a continuous form
- decreasing cost (versus time)

Disadvantages of carbon fibers include the following:

- anisotropy (in the axial versus transverse directions)
- low strain to failure
- compressive strength is low compared to tensile strength
- tendency to be oxidized and become a gas (e.g., CO) upon heating in air above about 400°C
- oxidation of carbon fibers is catalyzed by an alkaline environment

As each property is determined by the structure, the different properties are interrelated. The following trends usually go together:

- increase in the tensile modulus
- decrease in the strain to failure
- decrease in the compressive strength

- increase in the shear modulus
- increase in the degree of anisotropy
- decrease in the electrical resistivity
- increase in the thermal conductivity
- decrease in the coefficient of thermal expansion
- increase in density
- increase in thermal stability (oxidation resistance)
- increase in chemical stability
- increase in cost

Mechanical Properties

Table 4.1 [1] shows the tensile properties of carbon fibers along the fiber axis compared to those of a graphite single crystal along the *a*-axis, i.e., parallel to the carbon layers. Although the carbon layers in a carbon fiber exhibit a strong preferred orientation parallel to the fiber axis, the alignment of the layers is far from being perfect and the crystallite size is finite. Therefore, the tensile modulus and strength of carbon fibers are considerably below those of a graphite single crystal. The modulus of HM-type fibers approaches that of a graphite single crystal, but that of HT-type fibers is much below that of a graphite single crystal. The tensile strengths of both HM and HT fibers are very much below that of a graphite single crystal, although the strength of HT is higher than that of HM. There is thus much room for improvement of the tensile strength of carbon fibers. In contrast, there is not much room to improve the tensile modulus.

The tensile properties of some commercial carbon fibers of the high-performance (HP) grade are shown in Table 4.2 [2]. For the same precursor material (PAN or mesophase pitch), the tensile strength, modulus, and strain to failure vary over large ranges.

The tensile modulus is governed by the preferred orientation of the carbon layers along the fiber axis, so it increases with decreasing interlayer

Table 4.1 Considerations concerning Young's modulus (E) and the tensile strength (σ) of carbon fibers. From Ref. 1.

	Theoretical values for graphite single crystal	*Carbon fibers*		*Future trends*
		HT type	*HM type*	
Young's modulus, E	$E = 1\,000$ GPa	$E = 250$ GPa	$E = 700$ GPa	Further increase not necessary
Tensile strength, σ	$\sigma_{theor.} = E/10$ $= 100$ GPa	$\sigma_{theor.} = 25$ GPa $\sigma_{exp.} = 5$ GPa $= 20\%$ of $\sigma_{ther.}$	$\sigma_{theor.} = 70$ GPa $\sigma_{exp.} = 3$ GPa $= 4\%$ of $\sigma_{theor.}$	Further improvement expected

Table 4.2   Tensile modulus, strength, and strain to failure of carbon fibers. From Ref. 2.

Manufacturer	Fiber	Modulus (GPa)	Strength (GPa)	Strain to failure (%)
PAN-based, high modulus (low strain to failure)				
Celanese	Celion GY-70	517	1.86	0.4
Hercules	HM-S Magnamite	345	2.21	0.6
Hysol Grafil	Grafil HM	370	2.75	0.7
Toray	M50	500	2.50	0.5
PAN-based, intermediate modulus (intermediate strain to failure)				
Celanese	Celion 1000	234	3.24	1.4
Hercules	IM-6	276	4.40	1.4
Hysol Grafil	Apollo IM 43-600	300	4.00	1.3
Toho Beslon	Sta-grade Besfight	240	3.73	1.6
Amoco	Thornel 300	230	3.10	1.3
PAN-based, high strain to failure				
Celanese	Celion ST	235	4.34	1.8
Hercules	AS-6	241	4.14	1.7
Hysol Grafil	Apollo HS 38-750	260	5.00	1.9
Toray	T 800	300	5.70	1.9
Mesophase pitch-based				
Amoco	Thornel P-25	140	1.40	1.0
	P-55	380	2.10	0.5
	P-75	500	2.00	0.4
	P-100	690	2.20	0.3
	P-120	820	2.20	0.2

spacing (d_{002}) and with increasing L_c and L_a, as shown in Table 4.3 [3] for a series of mesophase pitch-based carbon fibers produced by du Pont.

Comparison of the du Pont fibers (Table 4.3) with the Amoco fibers (Table 4.2), both of which are based on mesophase pitch, indicates the superior tensile strength of the du Pont fibers. Unfortunately the du Pont fibers are not commercially available, whereas the Amoco fibers are.

Figure 4.1 [4] shows the tensile stress–strain curves of carbon fibers with different values of the tensile modulus. For a high-modulus carbon fiber (e.g., HM70), the stress–strain curve is a straight line up to failure; as the modulus decreases there is an increasing tendency for the slope to increase with increasing strain. This effect occurs because the fiber is increasingly stretched as the strain increases; the carbon layers become more aligned and the modulus therefore increases. It forms the basis of a process called stress-graphitization.

Table 4.3 Mechanical properties of pitch-based carbon fibers and their structural parameters as determined by X-ray diffraction: d_{002}, the interlayer spacing; L_c, the out-of-plane crystallite size; and L_a, the in-plane crystallite size parallel to the fiber axis. From Ref. 3.

Fibers	Tensile modulus (GPa)	Tensile strength (GPa)	d_{002} (nm)	L_c (nm)	L_a (nm)
E-35	241	2.8	0.3464	3.2	7.2
E-55	378	3.2	0.3430	8.2	16.2
E-75	516	3.1	0.3421	10.7	22.4
E-105	724	3.3	0.3411	17.3	46.1
E-120	827	3.4	0.3409	18.9	51.4
E-130	894	3.9	0.3380	24.0	180.4

The tensile strength is strongly influenced by flaws, so it increases with decreasing test (gage) length and with decreasing fiber diameter. Figure 4.2 shows the variation of the tensile strength with the fiber diameter for various PAN-based carbon fibers [5]. There are two types of flaws, namely surface flaws and internal flaws. The surface flaws control the strength of carbon fibers that have not been heat-treated above 1 000–1 200°C; the internal flaws control the strength of carbon fibers that have been heat-treated above 1 000–1 200°C [6]. Upon etching a fiber, the amount of surface flaws is decreased, causing the fiber strength to increase. The minimum practical gage length is 0.5 mm [7], even though the ultimate fragment length of a stressed single fiber composite is 0.3 mm [6] and it is the ultimate fragment length (also called the critical length) that determines the composite strength. Table 4.4 [7] shows the tensile

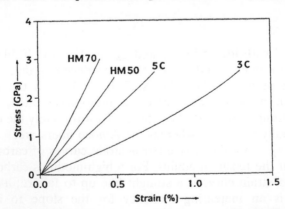

Figure 4.1 Tensile stress–strain curves of pitch-based carbon fibers (Carbonic HM50 and HM70) and PAN-based carbon fibers (Fortafil 3C and 5C). The test (gage) length is 100 mm. The strain rate is 1%/min. From Ref. 4. (Reprinted with permission from Pergamon Press Ltd.)

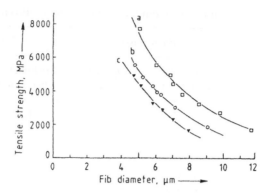

Figure 4.2 Relation between the tensile strength and fiber diameter: (a) Hercules AS-4 (Type HT), (b) Torayca T-300 (Type HT), and (c) Torayca M40 (Type HM). From Ref. 5.

strengths of carbon fibers (Hercules AS-4, PAN based) at different gage lengths, as determined by traditional tensile testing and by in situ fiber strength testing. The latter testing method involves embedding a single fiber in a matrix (e.g., epoxy) and pulling the unembedded ends of the fiber to increasing strain levels up to approximately three times greater than the failure strain of the fiber. While the strain is gradually increased, the number of breaks in the fiber is counted in situ [7]. The fiber eventually breaks into fragments of a length equal to the critical length, which is related to the tensile properties of the fiber and the interfacial shear strength between the fiber and the matrix. Table 4.4 shows that the fiber strength determined by either method increases with decreasing gage length. The latter method has the advantage of being

Table 4.4 Tensile strengths of AS-4 fibers at different gage lengths as determined by traditional tension testing and by in situ fiber strength testing in epoxy and solvent-deposited polycarbonate matrices. Tensile strengths appear in MPa followed by standard deviations. From Ref. 7.

Gage length (mm)	Conventional tension test (MPa)	In situ strength (MPa)	
		Epoxy	Polycarbonate
25.4	3 215 ± 966	3 188 ± 704	2 698 ± 518
8.0		3 850 ± 738	3 347 ± 725
4.0		4 264 ± 787	3 733 ± 752
2.0		4 720 ± 856	4 175 ± 773
1.0	5 285 ± 1 731	5 223 ± 911	4 582 ± 814
0.55	5 644 ± 994	5 693 ± 945	4 996 ± 869
0.3		6 189 ± 973	5 437 ± 883

Table 4.5 Tensile and compressive strength of carbon fibers. From Ref. 8.

	Pitch-based		PAN-based	
Carbon fiber	Carbonized fiber HTX	Graphitized fiber HMX	Carbonized fiber T-300	Graphitized fiber M40
Tensile strength $(\sigma_f)_{ten.}$ (GPa)	3.34	4.33	3.50	2.88
Estimated compressive strength $(\sigma_f)_{comp.}$ (GPa)	1.25	0.54	2.06	0.78
$(\sigma_f)_{comp.}/(\sigma_f)_{ten.}$ (%)	37.4	12.5	58.9	27.1

Table 4.6 Compressive failure strains for pitch-based fibers and PAN-based carbon fibers. From Ref. 9.

Precursor	Fiber	Modulus (GPa)	Mean failure strain (%)	Standard deviation (%)	No. of specimens
Pitch	HM	519	0.346	0.074	39
	HT	244	0.981	0.064	9
	UHM	662	0.163	0.063	9
	P-75	517	0.248	0.04	8
PAN	AS-1	228	2.59	0.181	8
	AS-4	241	2.27	0.256	8
	IM-6	276	2.35	0.132	8
	IM-7	303	2.11	0.317	8
	T-700	234	2.66	0.111	8
	T-300	230	2.36	0.125	8
	GY-30	241	2.65	0.207	8

applicable to very short gage lengths, but it has the disadvantage of being sensitive to the fiber prestrain resulting from the specimen preparation technique. The difference between the in situ fiber strengths for epoxy and polycarbonate matrices (Table 4.4) is due to a difference in the fiber prestrain.

The compressive strength is much lower than the tensile strength, as shown in Table 4.5 [8]. The ratio of the compressive strength to the tensile strength is smaller for graphitized fibers than carbonized fibers. Pitch-based carbon fibers have even lower compressive strength than PAN-based fibers. Moreover, the compressive failure strain is much lower for pitch-based carbon fibers than PAN-based carbon fibers, as shown in Table 4.6 [9]. These differences between pitch-based and PAN-based fibers are consistent with the

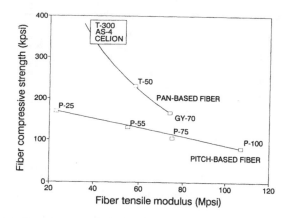

Figure 4.3   Relation between the compressive strength and tensile modulus of carbon fibers. From Ref. 11. (By permission of the Materials Research Society.)

difference in the compressive failure mechanism. Pitch-based fibers of high modulus typically deform by a shear mechanism, with kink bands formed on a fracture surface at 45° to the fiber axis. In contrast, PAN-based fibers typically buckle on compression and form kink bands at the innermost part of the fracture surface, which is normal to the fiber axis [10]. The difference in compressive behavior between pitch-based and PAN-based carbon fibers is attributed to the strong preferred orientation of the carbon layers in pitch-based fibers and the more random microstructure in PAN-based fibers. The oriented layer microstructure causes the fiber to be susceptible to shearing [9]. Thus, the compressive strength decreases with increasing tensile modulus, as shown in Figure 4.3 [11]. The axial Poisson's ratio of carbon fibers is around 0.26–0.28 [12].

The shear modulus of carbon fibers decreases with increasing L_c and with increasing L_a [4]. This is expected, since increases in L_c and L_a imply a greater degree of carbon layer preferred orientation. A decrease in the shear modulus is accompanied by a decrease in the compressive strength, as shown in Figure 4.4 [4]. The values of the shear modulus of various commercial carbon fibers are listed in Table 4.7 [13].

The values of the torsional modulus of various commercial carbon fibers are listed in Table 4.8 [6]. The torsional modulus is governed mostly by the cross-sectional microstructure. Mesophase pitch-based carbon fibers have low torsional modulus because they have an appreciable radial cross-sectional microstructure, which facilitates interlayer shear. Hence, the torsional modulus of mesophase pitch-based carbon fibers is even lower than that of isotropic pitch-based carbon fibers. On the other hand, PAN-based carbon fibers have high torsional modulus because they have an appreciable degree of circumferential microstructure [6].

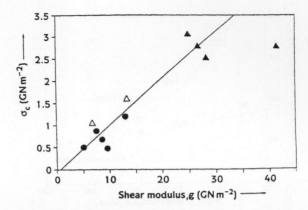

Figure 4.4 Relation between the compressive strength (σ_c) and the shear modulus (g) of carbon fibers. From Ref. 4. (Reprinted with permission from Pergamon Press Ltd.)

Table 4.7 Properties of some PAN-based carbon fibers. From Ref. 13.

Fiber	Tensile modulus (GPa)	Tensile strength (GPa)	Shear modulus (GPa)
T-300	230	3.5	17.0
M30	290	3.9	17.0
T-50	390	2.4	
M-40	400	2.7	15.8
M-46	450	2.35	14.8
GY-70	520	1.8	
T-800H	290	5.6	18.1
M-40J	390	4.3	17.0
M-46J	450	4.2	17.0
M-60J	585	3.8	

Electrical Properties

The electrical resistivity of the mesophase pitch-based carbon fibers of Table 4.3 is shown as a function of temperature from 2 to 300 K in Figure 4.5 [3]. The resistivity decreases with increasing temperature for each type of fiber. This is because the carrier density increases with temperature, just as for carbons and graphites in general. At a given test temperature, the resistivity decreases with increasing tensile modulus. This is because an increase in the tensile modulus is accompanied by a decrease in the concentration of defects, and defects cause carrier scattering.

Table 4.8 Torsional modulus and Young's modulus of various carbon fiber types. From Ref. 6.

Carbon fiber type	Young's modulus (GPa)	Torsional modulus (GPa)
Pitch mesophase		
PM-A	184	13.5
PM-B	262	9.0
		9.2
PM-C	364	10.8
		9.7
PM-C (1700)	400	10.4
Isotropic pitch		
KCF-200	80	16.4
KCF-2700	~80	18.0
PAN		
T-400	226	21.4
AS	215	21.0
HM-S(H)	~370	20.2
Modmor I	~400	28.2
HM-S(C)	~380	35.3
Rayon		
T-11 (carbon)	79	15.3
T-12 (graphite)	83	22.8
T-25	176	12.3
T-70	390	12.3
T-75	540	13.8
T-100	680	16.9

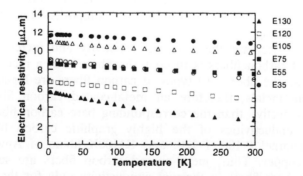

Figure 4.5 Variation of the electrical resistivity with temperature for the carbon fibers of Table 4.2. From Ref. 3. (By permission of IOP Publishing Limited.)

Table 4.9 Effect of nickel coating on the properties of PAN-based AS-4 (Hercules) carbon fiber. From Ref. 16.

Property	Bare fiber	Ni-coated fiber
Diameter (μm)	7.0	7.8
Density (g/cm^3)	1.80	2.97
Electrical resistivity ($10^{-6}\Omega$.cm)	1 530	7
Thermal conductivity (W/m/K)	7.2	10.7
Thermal expansion coefficient (10^{-6}/°C)	−1.7	−0.8
Tensile modulus (GPa)	234	210
Tensile strength (MPa)	3 582	2 582
Tensile elongation (%)	1.53	1.33

An effective way to decrease the resistivity of carbon fibers by a factor of up to 10 is intercalation. Intercalation is the formation of layered compounds in which foreign atoms (called the intercalate) are inserted between the carbon layers. The intercalate acts as an electron acceptor or an electron donor, thus doping the carbon fibers. This doping causes the carrier concentration to increase, thereby decreasing the electrical resistivity. Intercalation is only possible in relatively graphitic carbon fibers. For example, bromination (i.e., intercalation with bromine, an acceptor) causes a weight uptake of 18–20% for Amoco's Thornel P-100 and P-75 fibers, but 0% for P-55; it causes a resistivity decrease of 73–79% for P-120, P-100, and P-75 fibers, but just 4% for P-55 [14]. For the case of brominated P-100-4 (P-100-4 is even more graphitic than P-100), a resistivity of 11.0 $\mu\Omega$.cm has been reported [6]. On the other hand, the severity of the intercalation reaction in highly graphitic fibers can cause physical damage to the fibers, so that the mechanical properties and oxidation resistance are degraded [15]. Therefore, there is an optimum degree of graphitization, which corresponds to that of P-100 fibers for the case of bromine as the intercalate [14]. The low electrical resistivity of intercalated carbon fibers makes these fibers useful in composites for electromagnetic interference shielding.

A way to decrease the electrical resistivity and increase the thermal conductivity of carbon fibers is to coat the fibers with a metal that is more conductive than the fibers. All types of carbon fibers are higher in electrical resistivity than metals, therefore all metal-coated carbon fibers are more electrically conductive than the corresponding bare carbon fibers. However, the thermal conductivities of the highly graphitic carbon fibers, such as pitch-based Thornel P-100, P-120, and K1100X fibers of Amoco, are even higher than copper. Thus, metal-coated carbon fibers are superior to the corresponding bare fibers in thermal conductivity only for the less graphitic fibers, which constitute the vast majority of carbon fibers used in practice anyway. Table 4.9 [16] shows the effect of a 0.35 μm thick electrodeposited

Table 4.10   Thermal properties of the most advanced pitch-based carbon fibers. From Ref. 17.

Material	Longitudinal thermal conductivity (W/m/K)	CTE (10^{-6}/K)	Density (g/cm³)	Specific conductivity (W.cm³/m/K/g)
Al-6063	218	23	2.7	81
Copper	400	17	8.9	45
P-100	520	−1.6	2.2	236
P-120	640	−1.6	2.1	305
K1100X	1 100	−1.6	2.2	500
K1100X/Al (55 vol.%)	634	0.5	2.5	236
K1100X/epoxy (60 vol.%)	627	−1.4	1.8	344
K1100X/Cu (46 vol.%)	709	1.1	5.9	117
K1100X/C (53 vol.%)	696	−1.0	1.8	387

nickel coating on the properties of an originally 7.0 μm thick PAN-based AS-4 carbon fiber (Hercules), which is not graphitic. The nickel coating causes the density to increase, the electrical resistivity to decrease, the thermal conductivity to increase, the thermal expansion coefficient to increase, and the tensile modulus, strength, and elongation to decrease. The largest effect is the decrease in the electrical resistivity.

Thermal Conductivity

The longitudinal thermal conductivity, thermal expansion coefficient, density, and specific thermal conductivity (conductivity/density) of Amoco's mesophase pitch-based carbon fibers are shown in Table 4.10 [17]. The thermal conductivities of P-100, P-120, and K1100X fibers are all higher than that of copper, while the thermal expansion coefficients and densities are much lower than those of copper. Thus, the specific thermal conductivity is exceptionally high for these carbon fibers. In general, the thermal expansion coefficient of carbon fibers decreases with increasing tensile modulus, as shown in Figure 4.6 [18]. Table 4.10 also shows that use of K1100X fibers in Al, epoxy, Cu and C matrices results in composites of high thermal conductivity.

Oxidation Resistance

The oxidation resistance of carbon fibers increases with the degree of graphitization. Figure 4.7 shows the percentage weight remaining as a function of temperature during exposure of carbon fibers to flowing air for three types

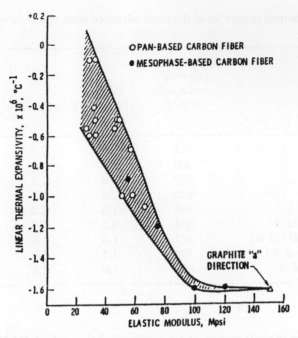

Figure 4.6 Relation between the longitudinal thermal expansion coefficient and the tensile modulus of carbon fibers. From Ref. 18.

carbon fiber (Thornel), which is graphitizable. As the degree of graphitization increases, the electrical resistivity decreases, the thermal conductivity increases, the thermal expansion coefficient decreases, and the tensile modulus, strength, and elongation to decrease. The largest effect is the decrease in the electrical resistivity.

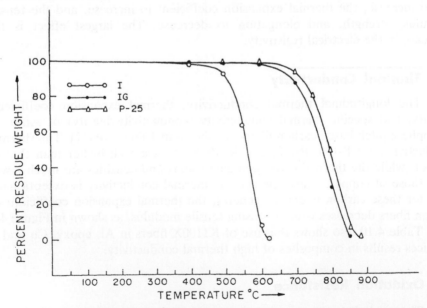

Figure 4.7 Percent residual weight during heating (in flow air) of three types of pitch-based carbon fibers. From Ref. 19. (Reprinted with permission from Pergamon Press Ltd.)

of fibers, namely I (Kureha isotropic pitch-based carbon fibers), IG (Kureha isotropic pitch-based carbon fibers graphitized at 2 700°C) and P-25 (Amoco mesophase pitch-based carbon fibers) [19]. In the case of I fibers, weight loss starts at about 400°C and is gradual up to 500°C; from 500 to 600°C, the weight loss is sharp; complete weight loss takes place at 620°C. In the case of IG fibers, weight loss starts at about 400°C, but is gradual up to 700°C; beyond 700°C, the weight loss is sharp; complete weight loss takes place by 850°C. In the case of P-25 fibers, weight loss starts at about 500°C and is only about 2% up to 600°C; complete weight loss takes place at about 880°C. The apparent activation energies range from 112 to 205 kJ/mol for various pitch-based carbon fibers [19]. The oxidation has been modeled to yield information on the oxidation kinetics, the activation energy, and the rate-determining step [20].

Severe oxidation causes carbon fibers to lose weight due to the evolution of CO or CO_2 gases. However, slight oxidation may cause carbon fibers to gain weight slightly due to the formation of chemical bonds to various oxygen-containing functional groups on the surface of the fibers. The oxygen-containing groups (or adsorbed oxygen) increase the polar component of the surface free energy and hence result in enhanced electrochemical response (relevant for fiber electrodes) and improved fiber–matrix bonding (when the fibers are used in a composite).

Although oxidation due to molecular oxygen is most common on earth, oxidation due to atomic oxygen is important in aerospace applications. Atomic oxygen is more reactive than molecular oxygen and is the dominant atmospheric element in low earth orbit [21].

Biocompatibility

Carbon is more biocompatible than even gold or platinum, so carbon fibers are used as implants, which act as a scaffold for collagen in tendons [22], for the repair of abdominal wall defects [23], and for the growth of spinal axons in a spinal cord [24]. The carbon fibers provide a favorable adhesive surface and a possible guiding function [24].

References

1. E. Fitzer and F. Kunkele, *High Temp. – High Pressures* **22**(3), 239–266 (1990).
2. D.J. Johnson, *J. Phys. D: Appl. Phys.* **20**(3), 286–291 (1987).
3. B. Nysten, J.-P. Issi, R. Barton, Jr., D.R. Boyington, and J.G. Lavin, *J. Phys. D: Appl. Phys.* **24**(5), 714–718 (1991).
4. M.G. Northolt, L.H. Veldhuizen, and H. Jansen, *Carbon* **29**(8), 1267–1279 (1991).
5. E. Fitzer and W. Frohs, *Chem. Eng. Technol.* **13**(1), 41–49 (1990).
6. J.-B. Donnet and R.C. Bansal, *International Fiber Science and Technology 10 (Carbon Fibers)*, 2d ed., Marcel Dekker, New York, 1990, pp. 267–366.
7. M.C. Waterbury and L.T. Drzal, *J. Compos. Technol. Res.* **13**(1), 22–28 (1991).

8. T. Ohsawa, M. Miwa, M. Kawade and E. Tsushima, *J. Appl. Polym. Sci.* **39**(8), 1733–1743 (1990).
9. J.M. Prandy and H.T. Hahn, in *Proc. Int. SAMPE Symp. and Exhib.*, *35, Advanced Materials: Challenge Next Decade*, edited by G. Janicki, V. Bailey, and H. Schjelderup, 1990, pp. 1657–1670.
10. D.J. Johnson, in *Carbon Fibers Filaments and Composites*, edited by J.L. Figueiredo, C.A. Bernardo, R.T.K. Baker, and K.J. Huttinger, Kluwer Academic, Dordrecht, 1990, pp. 119–146.
11. S. Kumar and T.E. Helminiak, *Mater. Res. Soc. Symp. Proc., Vol. 134 (Mater. Sci. Eng. Rigid-Rod Polym.*, 1989, 363–374.
12. I. Krucinska and T. Stypka, *Compos. Sci. Technol.* **41**, 1–12 (1991).
13. S. Kumar, in *Proc. Int. SAMPE Symp. and Exhib.*, *35, Advanced Materials: Challenge Next Decade*, edited by G. Janicki, V. Bailey, and H. Schjelderup, 1990, pp. 2224–2235.
14. D.D.L. Chung, *Ext. Abstr. Program Bienn. Conf. Carbon* **20**, 694–695 (1991).
15. C.T. Ho and D.D.L. Chung, *Carbon* **28**(6), 831–837 (1990).
16. R.E. Evans, D.E. Hall, and B.A. Luxon, *SAMPE Q.* **17**(4), 18–26 (1986).
17. Amoco Performance Products, data sheet, April 30, 1992.
18. W. de la Torre, in *Proc. 6th Int. SAMPE Electron. Conf.*, 1992, pp. 720–733.
19. T.L. Dhami, L.M. Manocha, and O.P. Bahl, *Carbon* **29**(1), 51–60 (1991).
20. I.M.K. Ismail and W.C. Hurley, *Carbon* **30**(3), 419–427 (1992).
21. M. Tagawa, N. Ohmae, M. Umeno, A. Yasukawa, K. Gotoh, and M. Tagawa, *Jpn. J. Appl. Phys.* **30**(9), 2134–2138 (1991).
22. D.H.R. Jenkins and B. McKibbin, *J. Bone Joint Surg.* **62B**(4), 497–499 (1980).
23. A. Cameron and D. Taylor, in *Proc. Institute of Basic Medical Sciences Symp. on Interaction of Cells with Natural and Foreign Surfaces, Royal College of Surgeons of England, 1984*, Plenum Press, New York, 1986, pp. 271–278.
24. T. Khan, M. Dauzvardis, and S. Sayers, *Brain Research* **541**(1), 139–145 (1991).

PART **II**

Carbon Fiber Composites

CHAPTER 5

Introduction to Carbon Fiber Composites

Composite materials refer to materials containing more than one phase such that the different phases are artificially blended together. They are not multiphase materials in which the different phases are formed naturally by reactions, phase transformations, or other phenomena.

A composite material typically consists of one or more fillers in a certain matrix. A carbon fiber composite refers to a composite in which at least one of the fillers is carbon fibers, either short or continuous, unidirectional or multidirectional, woven or nonwoven. The matrix is usually a polymer, a metal, a carbon, a ceramic, or a combination of different materials. Except for sandwich composites, the matrix is three-dimensionally continuous, whereas the filler can be three-dimensionally discontinuous or continuous. Carbon fiber fillers are usually three-dimensionally discontinuous, unless the fibers are three-dimensionally interconnected by weaving or by the use of a binder such as carbon.

The high strength and modulus of carbon fibers makes them useful as a reinforcement for polymers, metals, carbons, and ceramics, even though they are brittle. Effective reinforcement requires good bonding between the fibers and the matrix, especially for short fibers. For a unidirectional composite (i.e., one containing continuous fibers all in the same direction), the longitudinal tensile strength is quite independent of the fiber–matrix bonding, but the transverse tensile strength and the flexural strength (for bending in longitudinal or transverse directions) increase with increasing fiber–matrix bonding. On the other hand, excessive fiber–matrix bonding can cause a composite with a brittle matrix (e.g., carbon and ceramics) to become more brittle, as the strong fiber–matrix bonding causes cracks to propagate straightly, in the direction perpendicular to the fiber–matrix interface without being deflected to propagate along this interface. In the case of a composite with a ductile matrix (e.g., metals and polymers), a crack initiating in the brittle fiber tends to be blunted when it reaches the ductile matrix, even when the fiber–matrix bonding is strong. Therefore, an optimum degree of fiber–matrix bonding is needed for brittle-matrix composites, whereas a high degree of fiber–matrix bonding is preferred for ductile-matrix composites.

The mechanisms of fiber–matrix bonding include chemical bonding, van der Waals bonding, and mechanical interlocking. Chemical bonding gives the largest bonding force, provided that the density of chemical bonds across the fiber–matrix interface is sufficiently high. This density can be increased by chemical treatments of the fibers or by sizings on the fibers. Mechanical interlocking between the fibers and the matrix is an important contribution to the bonding if the fibers form a three-dimensional network. Otherwise, the fibers should have a rough surface in order for a small degree of mechanical interlocking to take place.

Both chemical bonding and van der Waals bonding require the fibers to be in intimate contact with the matrix. For intimate contact to take place, the matrix or matrix precursor must be able to wet the surfaces of the carbon fibers during infiltration of the matrix or matrix precursor into the carbon fiber preform. Chemical treatments and coatings can be applied to the fibers to enhance wetting. The choice of treatment or coating depends on the matrix. Another way to enhance wetting is the use of a high pressure during infiltration. A third method is to add a wetting agent to the matrix or matrix precursor before infiltration. As the wettability may vary with temperature, the infiltration temperature can be chosen to enhance wetting.

The occurrence of a reaction between the fibers and the matrix helps the wetting and bonding between the fibers and the matrix. However, an excessive reaction degrades the fibers, and the reaction product(s) may be undesirable for the mechanical, thermal, or moisture resistance properties of the composite. Therefore, an optimum amount of reaction is preferred.

Carbon fibers are electrically and thermally conductive, in contrast to the nonconducting nature of polymer and ceramic matrices. Therefore, carbon fibers can serve not only as a reinforcement, but also as an additive for enhancing the electrical or thermal conductivity. Furthermore, carbon fibers have nearly zero coefficient of thermal expansion, so they can also serve as an additive for lowering the thermal expansion. The combination of high thermal conductivity and low thermal expansion makes carbon fiber composites useful for heat sinks in electronics and for space structures that require dimensional stability. As the thermal conductivity of carbon fibers increases with the degree of graphitization, applications requiring a high thermal conductivity should use the graphitic fibers, such as the high-modulus pitch-based fibers and the vapor grown carbon fibers. Carbon fibers are more cathodic than practically any metal, so in a metal matrix, a galvanic couple is formed with the metal as the anode. This causes corrosion of the metal. The corrosion product tends to be unstable in moisture and causes pitting, which aggravates corrosion. To alleviate this problem, carbon fiber metal-matrix composites are often coated.

Carbon is the matrix that is most compatible to carbon fibers. The carbon fibers in a carbon-matrix composite (called carbon–carbon composite) serve to strengthen the composite, as the carbon fibers are much stronger than the carbon matrix due to the crystallographic texture in each fiber. Moreover, the carbon fibers serve to toughen the composite, as the debonding between the fibers and the matrix provides a mechanism for energy absorption during mechanical deformation. In addition to having attractive mechanical properties, carbon–carbon composites are more thermally conductive than carbon fiber polymer-matrix composites. However, at elevated temperatures (above 320°C), carbon–carbon composites degrade due to the oxidation of carbon, (especially the carbon matrix), which forms CO_2 gas. To alleviate this problem, carbon–carbon composites are coated.

Carbon fiber ceramic-matrix composites are more oxidation resistant than carbon–carbon composites. The most common form of such composites is carbon fiber reinforced concrete. Although the oxidation of carbon is catalyzed by an alkaline environment and concrete is alkaline, the chemical stability of carbon fibers in concrete is superior to that of competitive fibers, such as polypropylene, glass, and steel. Composites containing carbon fibers in more advanced ceramic matrices (such as SiC) are rapidly being developed.

Carbon fiber composites are most commonly fabricated by the impregnation (or infiltration) of the matrix or matrix precursor in the liquid state into the fiber preform, which is most commonly in the form of a woven fabric. In the case of composites in the shape of tubes, the fibers may be impregnated in the form of a continuous bundle from a spool and subsequently the bundles may be wound on a mandrel. Instead of impregnation, the fibers and matrix material may be intermixed in the solid state by commingling carbon fibers and matrix fibers, by coating the carbon fibers with the matrix material, by sandwiching carbon fibers with foils of the matrix material, and in other ways. After impregnation or intermixing, consolidation is carried out, often under heat and pressure.

Due to the decreasing price of carbon fibers, the applications of carbon fiber composites are rapidly widening to include the aerospace, automobile, marine, construction, biomedical, and other industries. This situation poses an unusual demand on research and development in the field of carbon fiber composites.

CHAPTER 6

Polymer-Matrix Composites

Polymers

Polymer-matrix composites are much easier to fabricate than metal-matrix, carbon-matrix, and ceramic-matrix composites, whether the polymer is a thermoset or a thermoplast. This is because of the relatively low processing temperatures required for fabricating polymer-matrix composites. For thermosets, such as epoxy, phenolic, and furfuryl resin, the processing temperature typically ranges from room temperature to about 200°C; for thermoplasts, such as polyimide (PI), polyethersulfone (PES), polyetheretherketone (PEEK), polyetherimide (PEI), and polyphenyl sulfide (PPS), the processing temperature typically ranges from 300 to 400°C.

Thermosets (especially epoxy) have long been used as polymer matrices for carbon fiber composites. During curing, usually performed in the presence of heat and pressure, a thermoset resin hardens gradually due to the completion of polymerization and the cross-linking of the polymer molecules. Thermoplasts have recently become important because of their greater ductility and processing speed compared to thermosets, and the recent availability of thermoplasts that can withstand high temperatures. The higher processing speed of thermoplasts is due to the fact that thermoplasts soften immediately upon heating above the glass transition temperature (T_g) and the softened material can be shaped easily. Subsequent cooling completes the processing. In contrast, the curing of a thermoset resin is a reaction which occurs gradually.

Epoxy is by far the most widely used polymer matrix for carbon fibers. Trade names of epoxy include Epon, Epi-rez, and Araldite. Epoxy has an excellent combination of mechanical properties and corrosion resistance, is dimensionally stable, exhibits good adhesion, and is relatively inexpensive. Moreover, the low molecular weight of uncured epoxide resins in the liquid state results in exceptionally high molecular mobility during processing. This mobility enables the resin to quickly wet the surface of carbon fiber, for example.

Epoxy resins are characterized by having two or more epoxide groups per molecule. The chemical structure of an epoxide group is:

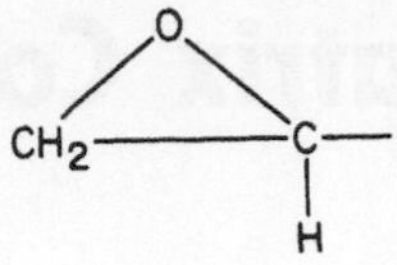

$$
\underset{\text{CH}_2}{}\overset{\displaystyle O}{\diagup\diagdown}\underset{\text{H}}{\overset{\displaystyle |}{\text{C}}}\text{—}
$$

Most commercial epoxy resins have the general chemical structure:

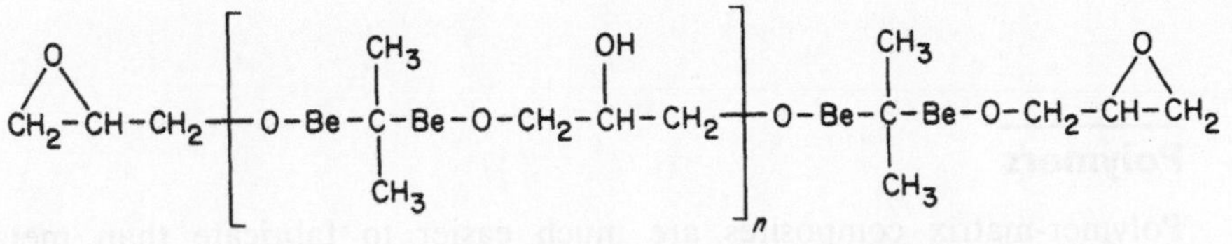

$$
\text{CH}_2\text{-CH-CH}_2 \left[\text{O-Be}-\underset{\text{CH}_3}{\overset{\text{CH}_3}{\text{C}}}-\text{Be-O-CH}_2-\overset{\text{OH}}{\text{CH}}-\text{CH}_2 \right]_n \text{O-Be}-\underset{\text{CH}_3}{\overset{\text{CH}_3}{\text{C}}}-\text{Be-O-CH}_2-\text{CH-CH}_2
$$

where Be = benzene ring. For liquids, n is usually less than 1; for solid resins, n is 2 or greater.

The curing of an epoxy resin requires a cross-linking agent and/or a catalyst. The epoxy and hydroxyl groups $(-\text{OH})$ are the reaction sites for cross-linking. Cross-linking agents include amines, anhydrides, and aldehyde condensation products. In the curing reaction, the epoxide ring is opened (called ring scission) and a donor hydrogen from, say, an amine or hydroxyl group bonds with the oxygen atom of the epoxide group. Ethylene diamine is an amine which serves as a cross-linking agent.

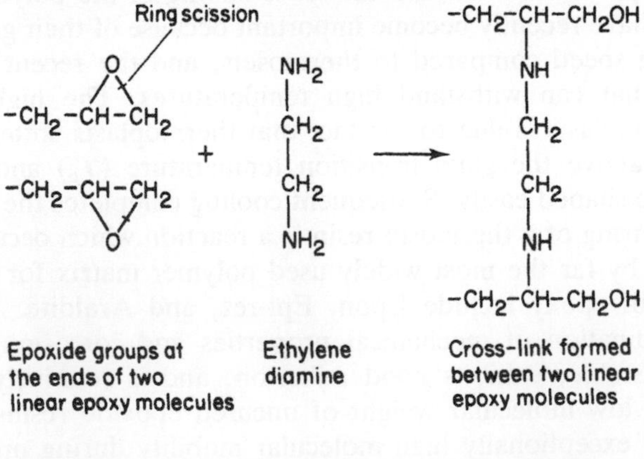

| Epoxide groups at the ends of two linear epoxy molecules | Ethylene diamine | Cross-link formed between two linear epoxy molecules |

As no by-product is given off during curing, shrinkage is low.

The mers (repeating units) of typical thermoplasts used for carbon fibers are shown below, where Be = benzene ring.

PI

PEEK   ...–O–Be–O–Be–C–Be–...

PPS   ...–Be–S–...

PES   ...–Be–O–Be–S–...

PEI

The properties of the above thermoplasts are listed in Table 6.1. In contrast, epoxies have tensile strengths of 30–100 MPa, moduli of elasticity of 2.8–3.4 GPa, ductilities of 0–6% and a density of 1.25 g/cm³ [3]. Thus, epoxies are much more brittle than PES, PEEK, and PEI. In general, the ductility of a semicrystalline thermoplast decreases with increasing crystallinity. For example, the ductility of PPS can range from 2 to 20%, depending on the

Table 6.1 Properties of thermoplasts for carbon fiber polymer-matrix composites.

	PES	*PEEK*	*PEI*	*PPS*	*PI*
T_g(°C)	230[a]	170[a]	225[a]	86[a]	256[b]
Decomposition temperature (°C)	550[a]	590[a]	555[a]	527[a]	550[b]
Processing temperature (°C)	350[a]	380[a]	350[a]	316[a]	304[b]
Tensile strength (MPa)	84[c]	70[c]	105[c]	66[c]	138[b]
Modulus of elasticity (GPa)	2.4[c]	3.8[c]	3.0[c]	3.3[c]	3.4[b]
Ductility (% elongation)	30–80[c]	50–150[c]	50–65[c]	2[c]	5[b]
Izod impact (ft lb/in.)	1.6[c]	1.6[c]	1[c]	<0.5[c]	1.5[c]
Density (g/cm³)	1.37[c]	1.31[c]	1.27[c]	1.3[c]	1.37[b]

[a]From Ref. 1.
[b]From Ref. 2.
[c]From Ref. 3.

crystallinity [4]. Another major difference between thermoplasts and epoxies lies in the higher processing temperatures of thermoplasts (300–400°C).

Much work has been done to improve epoxies for controlling the fiber–matrix interface [5,6], increasing the toughness [7,8], and reducing the moisture sensitivity [9]. Other than epoxies, thermosets used for carbon fibers include polyimide [10] and bismaleimide [11,12]. (Polyimides can be thermoplasts or thermosets.)

Semicrystalline thermoplasts (e.g., PEEK) are more efficiently reinforced than are amorphous thermoplasts (e.g., PES). This is because the fibers act as nucleation sites for crystallization; the fiber becomes surrounded by a microcrystalline structure, which binds the fiber more firmly to the polymer and improves the modulus. Furthermore, the degree of reinforcement increases with the degree of crystallinity [13].

The addition of fibers increases the softening temperature of a thermoplast and the effect is greater with semicrystalline polymers than with amorphous polymers, where the gain is typically 10–20°C. This is because softening is governed by T_g for an amorphous polymer, but is governed by the melting point (T_m) and the degree of crystallinity for a semicrystalline polymer [13].

The addition of fibers increases the creep resistance because it impedes the molecular mobility. The effect is greater with amorphous thermoplasts than with semicrystalline thermoplasts, as crystalline polymers themselves inhibit creep [13].

Water absorbed by a polymer acts as a plasticizer and decreases strength and stiffness, but increases toughness. As fibers absorb much less water than polymers, addition of fibers decreases the amount of water absorption [13]. It also increases the dimensional stability when the temperature is changed, because fibers have much lower thermal expansion coefficients than polymers [13].

The use of fibers produces higher melt viscosities at a given shear rate, so higher processing temperatures and/or higher injection pressures are necessary. On the other hand, the addition of fibers reduces shrinkage during processing [13].

Surface Treatments of Carbon Fibers for Polymer Matrices

Surface treatments of carbon fibers are essential for improving the bonding between the fibers and the polymer matrix. They involve oxidation treatments and the use of coupling agents, wetting agents, and/or sizings (coatings). Carbon fibers need treatment both for thermosets and thermoplasts. As the processing temperature is usually higher for thermoplasts than thermosets, the treatment must be stable to a higher temperature (300–400°C) when a thermoplast is used.

Oxidation treatments can be applied by gaseous, solution, electrochemical, and plasma (e.g., acid plasma [14]) methods. They serve mainly to

Table 6.2 Effects of various surface treatments on properties of high-modulus carbon fibers and their epoxy-matrix composites. From Ref. 15.

	Fiber properties		Composite properties	
Fiber treatment	Wt. loss (%)	Tensile strength loss (%)	Flexural strength loss (%)	ILSS gain (%)
400°C in air (30 min.)	0	0	0	18
500°C in air (30 min.)	0.4	6	12	50
600°C in air (30 min.)	4.5	50	Too weak to test	–
60% HNO_3 (15 min.)	0.2	0	8	11
5.25% NaOCl (30 min.)	0.4	1.5	5	30
10–15% NaOCl (15 min.)	0.2	0	8	6
15% $HClO_4$ (15 min.)	0.2	0	12	0
5% $KMnO_4$/10% NaOH (15 min.)	0.4	0	15	19
5% $KMnO_4$/10% H_2SO_4 (15 min.)	6.0(+)	17	13	95
10% H_2O_2/20% H_2SO_4 (15 min.)	0.1	5	14	0
42% HNO_3/30% H_2SO_4 (15 min.)	0.1	0	4(+)	0
10% $NaClO_3$/15% NaOH (15 min.)	0.2	0	12	12
10% $NaClO_3$/25% H_2SO_4 (15 min.)	0.2	2	5(+)	91
15% $NaClO_3$/40% H_2SO_4 (15 min.)	0.7	4	15	108
10% $Na_2Cr_2O_7$/25% H_2SO_4 (15 min.)	0.3	8	15(+)	18
15% $Na_2Cr_2O_7$/40% H_2SO_4 (15 min.)	1.7	27	31	18

All liquid treatments at reflux temperature.

remove a weak surface layer from the fibers [15]. More severe oxidation treatments also serve to roughen the fiber surface, thereby enhancing the mechanical interlocking between the fibers and the matrix [16,17]. Chemical modification (particularly the production of carbonyl, carboxyl, and hydroxyl groups on the fiber surface) occurs, but contributes little to the fiber–matrix adhesion. The oxidation treatments essentially do not improve the fiber wetting [15]. However, the effect of the treatments varies with the polymer species.

Table 6.2 [15] shows the effect of oxidation treatments on the mechanical properties of high-modulus carbon fibers and their epoxy-matrix composites. The treatments degrade the fiber properties but improve the composite properties. The most effective treatment in Table 6.2 is refluxing in a 10% $NaClO_3$/25% H_2SO_4 mixture for 15 min., as this treatment results in a fiber weight loss of 0.2%, a fiber tensile strength loss of 2%, a composite flexural strength gain of 5%, and a composite interlaminar shear strength (ILSS) gain of 91%. Epoxy embedded single fiber tensile testing showed that anodic oxidation of pitch-based carbon fibers in ammonium sulfate solutions increased the interfacial shear strength by 300% [18]. As the modulus of the fiber

Table 6.3 Interfacial shear strength of carbon fibers in an epoxy matrix. From Ref. 19.

Fiber	Interfacial shear strength (MPa)
AU-1	48
AS-1	74
AU-4	37
AS-4	61

increases, progressively longer treatment times are required to attain the same improvement in ILSS. Although the treatment increases ILSS, it decreases the impact strength (i.e., impact energy), so the treatment time must be carefully controlled in order to achieve a balance in properties. The choice of treatment time also depends on the particular fiber–resin combination used. For a particular treatment, as the modulus of the fiber increases, the treatment's positive effect on the ILSS and its negative effect on the impact strength both become more severe [15].

Commercial carbon fibers are surface treated to enhance the bonding with epoxy, though the surface treatment is proprietary. Table 6.3 [19] shows the effect of a surface treatment on the interfacial shear strength for PAN-based carbon fibers manufactured by Hercules. In Table 6.3, fibers designated AS-1 and AS-4 are typical Type II intermediate strain fibers, whereas AU-1 and AU-4 are the untreated analogs of AS-1 and AS-4, respectively. The interfacial shear strength was determined from the critical shear transfer length, i.e., the length of the fiber fragments after fracture of a single fiber pulled in tension while being encapsulated in epoxy. Surface treatment increased the interfacial shear strength by 54 and 65% for AS-1 and AS-4 fibers, respectively [19]. Table 6.4 [19] shows the atomic surface concentrations of AS-1 and AS-4 fibers, as determined by X-ray photoelectron spectroscopy. The atomic surface concentrations are similar for AS-1 and AS-4, indicating that the superior interfacial shear strength of AS-1 compared to AS-4 is not due to a difference in the surface composition. On the other hand, scanning electron microscopy shows that the surface of AS-1 is corrugated, whereas that of AS-4 is smooth. Therefore, the superior interfacial shear strength of AS-1 is attributed to its surface morphology, which increases its surface area and enhances the mechanical interlocking between the fiber and the matrix [19].

Although surface treatments of carbon fibers result in some degree of oxidation, which places oxygen on the surface in an acidic form, the treatments themselves produce little acidity. Surface acidification is not desirable because it is accompanied by surface degradation [20].

The oxidation treatments approximately double the surface concentration

Table 6.4 Atomic surface concentrations of the carbon fibers of Table 6.3. From Ref. 19.

Fiber	Atomic percent				
	C_{1s}	O_{1s}	N_{1s}	S_{2p}	Na_{KLL}
AS-1	84	11	4.3	0.2	1.0
AS-4	83	12	4.0	0.2	0.7

of oxygen. Functional groups on the fiber surface and at 500 Å below the surface are listed below in relative order of abundance.

The main functional groups produced are carbonyl, carboxyl, and hydroxyl [15].

The oxygen concentration does not determine or correlate with the increase in ILSS or transverse tensile strength of the composites [15,16]. Indeed, the addition of the surface chemical oxygen groups is believed to be responsible for only 10% of the increase in adhesion resulting from the treatment; only about 4% of the surface sites of the carbon fibers are involved in chemical bonding with the epoxy and amine groups of the polymer. Although the magnitude of the bond strength for chemical bonds is very high, the quantity of bonds is low [21]. On the other hand, elimination of the functional groups on the treated fibers by diazomethane causes the ILSS to decrease toward the level of the untreated fibers [22], so the contribution of the functional groups to fiber–matrix adhesion cannot be neglected.

In addition to oxidation treatments, carbon fibers require the use of coupling agents, wetting agents, and/or sizings (coatings) in order to improve the wetting of the fibers by the polymer, the adhesion between the fibers and the matrix, and the handleability of the fibers. As one agent often serves more than one function, the distinction among coupling agents, wetting agents, and sizings is often vague.

Coupling agents [15,23] are mostly short-chain hydrocarbon molecules, one end of which is compatible or interacts with the polymer while the other end interacts with the fiber. A coupling agent molecule has the form $X-R$, where X interacts with the fiber and R is compatible or interacts with the polymer.

Organosilanes are of the form $R-Si-(OX)_3$, where X is methyl, ethyl, methoxyethyl, etc., and R is a suitable hydrocarbon chain. They are widely used as coupling agents between glass fibers and thermosets, as the $-OX$ groups react with the $-OH$ groups on the glass surface.

$$
\begin{array}{c}
| \\
O \\
| \\
Si-OH + XO-R \longrightarrow Si-O-R + HOX \\
| \\
O \\
|
\end{array}
$$

However, organosilanes do not function for carbon, organic, or metallic fibers.

Organotitanates and organozirconates function as coupling agents for both siliceous (e.g., glass) and nonsiliceous (e.g., carbon) fillers. The general formula of the organotitanates is:

$$
X-O-Ti \left[O-P-O-P-(OR)_2 \right]_3
$$

where R is usually a short-chain hydrocarbon such as C_8H_{17} and X is a group capable of interacting with the fiber. Organotitanates and organozirconates in amounts of 0.1–0.5 wt.% of formulation solids provide improved bonding between carbon fibers and thermosets (e.g., epoxy, polyurethane, polyester, and vinyl ester resins) [24].

Wetting agents are polar molecules with one end attracted to the fiber and the other end to the polymer. The agent forms a protective layer around the fiber, thereby improving dispersion. It also promotes adhesion by allowing more efficient wetting of the polymer on the fiber. The main difference between a wetting agent and a coupling agent is that a coupling agent forms a chemical bond with the fiber but a wetting agent does not [15].

Sizings (coatings or finishes) are commonly applied to carbon fibers in order to improve fiber–polymer adhesion and fiber handleability. The handleability is particularly important if the fibers are to be woven. The choice of sizing material depends on the polymer matrix. In particular, thermoplast-matrix composites require sizings that can withstand higher temperatures than thermoset-matrix composites, because of the higher processing temperature of the fiber. Sizing materials include prepolymers/polymers, carbon, SiC, and metals. Due to the relative ease of application, polymers (either cured or partially cured) are the most common sizing materials. Sizing thicknesses typically range from 0.1 to 1 μm.

Commercial carbon fibers are usually coated with a proprietary epoxy-compatible finish. Nevertheless, epoxy is the main sizing material for fibers used for epoxy-matrix composites. As the epoxy sizing decomposes at about 250°C, it is not very suitable for thermoplast-matrix composites, though it is still useful [25]. Instead, polyimides and polyimide–PES blends are used as sizings for carbon fibers in thermoplast-matrix composites. Polyimide-coated carbon fibers can withstand temperatures up to 450°C [1].

Other than epoxy, a number of polymers have been used as sizings for carbon fibers in epoxy-matrix composites. They include polyhydroxyether, polyphenyleneoxide, copolymers of styrene and maleic anhydride (SMA), a block copolymer of SMA with isoprene, polysulfone, polybutadiene, silicone, a carboxy-terminated polybutadiene/acrylonitrile copolymer, a copolymer of maleic anhydride and butadiene, and a copolymer of ethylene and acrylic acid [26]. In particular, an SMA coating results in a 50% increase in the interfacial shear strength (measured by using a single fiber) compared with commercially treated fibers, while causing no degradation in the impact strength [27]. In contrast, elastomer coatings result in improved crack resistance and impact strength [28].

Different methods are used to coat carbon fibers with polymers, namely deposition from solution, electrodeposition, and electropolymerization. Polymer deposition from solution was the most common, though recent work employs mostly electrodeposition or electropolymerization. An example of deposition from solution is the deposition of polyhydroxyether from a solution containing 0.91 wt.% of polyhydroxyether in Cellosolve. Polyhydroxyether is:

$$\left[-O-Be-\underset{\underset{CH_3}{|}}{\overset{\overset{CH_3}{|}}{C}}-Be-O-CH_2-\underset{\underset{OH}{|}}{\overset{\overset{H}{|}}{C}}-CH_2- \right]_n$$

The resulting polyhydroxyether coating increased the ILSS and flexural strength of the carbon fiber epoxy-matrix composite by 81% and 14%, respectively [25]. In electrodeposition, preformed polymers carrying ionized groups migrate to the oppositely charged electrode under an applied voltage. Electropolymerization involves the polymerization of monomers in an electrolytic cell. Solvents such as dimethyl formamide and dimethyl sulfoxide proved suitable. Since carbon fibers are electrically conducting, they serve as a good substrate for these electrical coating methods, which have the advantage

Table 6.5 Effect of oxidation treatment and polymer coating on ILSS of composite. From Ref. 22.

Oxidation	Polymer coating	Polymer (%)	Density (g/cm³)	ILSS (MPa)
None	None	–	1.28	16.2
60% HNO₃, 24 h.	None	–	1.29	24.3
	PVA	7	1.31	42.8
	PVC	7	1.31	42.1
	Rigid polyurethane	3	1.27	40.7
	PAN	7	1.27	16.6

of yielding uniform layers of readily controlled thickness in just a short time [29]. Sometimes grafting of the polymer to the fiber surface takes place, as in the case of the electrochemical oxidation of ω-diamines on carbon fibers, where the grafting provides a continuous succession of covalent bonds from the carbon fiber surface to the epoxy resin [30]. Deposition techniques have given the greatest improvement in composite properties—up to 60% in impact strength, 84% in work to fracture, and 90% in ILSS, but not simultaneously [26]. Table 6.5 [21] shows the effect of oxidation treatment and polymer coating on ILSS. The use of both oxidation treatment and polymer coating give the highest ILSS.

The sizing increases the ILSS; this changes the mode of composite fracture from growth of an interfacial crack to growth of a crack perpendicular to the fiber axis [15]. In some cases, the epoxy matrix penetrates the polymer coating [26]. An interphase between the fiber and the epoxy matrix is believed to exist. It is a three-dimensional region including not only the two-dimensional fiber–matrix interface, but also regions on both sides of the interface [21,31–33].

The polymer sizings are chemically bonded to the carbon fiber via its functional groups. In the case of an epoxy sizing, the epoxy group and amine group can react with the functional groups on the fiber surface [21]. Furthermore, the functional groups act as a catalyst for cross-linking, if their concentrations are not too high [29]. In the case of a polyimide sizing, the carboxylic acid groups in polyimide precursors react with the functional groups on the fiber surface [1].

Far less common than polymer coatings are carbon, SiC, and metal coatings on carbon fibers. Carbon coatings deposited by using acetylene vapor increase the ILSS from 34 MPa at 0 wt.% C coating to 56 MPa at 22 wt.% C coating [22]. SiC coatings, in the form of β-SiC single crystal whiskers grown on the carbon fiber surface perpendicular to the fiber axis, significantly increase the ILSS [22]. Metal (Ni,Cu) coatings deposited by electroless plating or electroplating on carbon fibers provide polar surfaces due to the presence of oxides and hydration of the surface [34].

Classification

Carbon fiber polymer-matrix composites can be classified according to whether the matrix is a thermoset or a thermoplast. Thermoset-matrix composites are by tradition far more common, but thermoplast-matrix composites are under rapid development. The advantages of thermoplast-matrix composites compared to thermoset-matrix composites include the following:

Lower manufacturing cost

- no cure
- unlimited shelf-life
- reprocessing possible (for repair and recycling)
- less health risks due to chemicals during processing
- low moisture content
- thermal shaping possible
- weldability (fusion bonding possible)

Better performance

- high toughness (damage tolerance)
- good hot/wet properties
- high environmental tolerance

The disadvantages of thermoplast-matrix composites include the following:

- limitations in processing methods
- high processing temperatures
- high viscosities
- stiff and dry prepregs when a solvent is not used (i.e., not drapeable or tacky)
- fiber surface treatments less developed

Carbon fiber polymer-matrix composites can be classified according to whether the fibers are short or continuous. Continuous fibers have much more effect than short fibers on the composite's mechanical properties, electrical resistivity, thermal conductivity, and on other properties, too. However, they give rise to composites that are more anisotropic. Continuous fibers can be in unidirectionally aligned tape or woven fabric form.

Fabrication

Short-fiber composites are usually fabricated by mixing the fibers with a liquid resin to form a slurry then molding to form a composite. The liquid resin is the unpolymerized matrix material in the case of a thermoset; it is the molten polymer or the polymer dissolved in a solvent in the case of a thermoplast. The molding methods are those conventionally used for polymers by themselves. For thermoplasts, the methods include injection molding (heating above the melting temperature of the thermoplast and forcing the slurry into a closed die by a plunger or a screw mechanism), extrusion (forcing the slurry through a die opening by using a screw mechanism), calendering (pouring the slurry into a set of rollers with a small opening between adjacent rollers to form a thin sheet), and thermoforming (heating above the softening temperature of the thermoplast and forming over a die (using matching dies, a vacuum, or air pressure), or without a die (using movable rollers)). For thermosets, compression molding or matched die molding (applying a high pressure and temperature to the slurry in a die to harden the thermoset) is commonly used. The casting of the slurry into a mold is not usually suitable because the difference in density between the resin and the fibers causes the fibers to float or sink, unless the viscosity of the resin is carefully adjusted. For forming a short-fiber composite coating, the fiber–resin slurry can be sprayed instead of molded.

Instead of using a fiber–resin slurry, short carbon fibers in the form of a mat or a continuous spun staple yarn can be impregnated with a resin and shaped using methods commonly used for continuous fiber composites. By using spun staple yarns from the Heltra process, researchers have produced epoxy-matrix composites that retain 97% of the tensile modulus and 70% of the tensile strength of their counterparts containing continuous carbon fiber tows. This is in spite of the discontinuity and slight twist in the fibers of the staple yarns [35].

Yet another method involves using continuous staple yarns in the form of an intimate blend of short carbon fibers and short thermoplast fibers. The yarns may be woven, if desired. They do not need to be impregnated with a resin to form a composite, as the thermoplast fibers melt during consolidation under heat and pressure [35].

Since carbon fibers are electrically conducting, they can be aligned in an electric field. Short (1.6 cm) carbon fibers in a nylon matrix have been aligned in this way, such that on average 68% of the fibers were within ±10° of perfect alignment [36].

Continuous fiber composites are commonly fabricated by hand lay-up of unidirectional fiber tapes or woven fabrics and impregnation with a resin. The molding, called bag molding, is done by placing the tapes or fabrics in a die and introducing high-pressure gases or a vacuum via a bag to force the individual plies together. Bag molding is widely used to fabricate large composite components for the skins of aircraft.

A method for forming unidirectional fiber composite parts with a constant cross section (e.g., round, rectangular, pipe, plate, I-shaped) is pultrusion, in which fibers are drawn from spools, passed through a polymer resin bath for impregnation, and gathered together to produce a particular shape before entering a heated die (Figure 6.1) [37].

A method for forming continuous carbon fiber composites of intricate

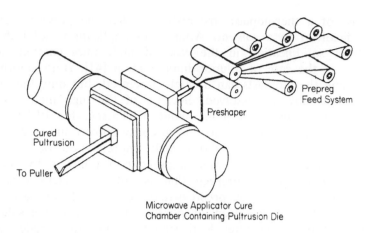

Preshaper

Prepreg
Feed System

Cured
Pultrusion

To Puller

Microwave Applicator Cure
Chamber Containing Pultrusion Die

Figure 6.1   Schematic of the pultrusion process. From Ref. 37. (Reprinted by permission of Kluwer Academic Publishers.)

shapes is resin transfer molding (RTM), in which a carbon fiber preform (usually prepared by braiding and held under compression in a mold) is impregnated with a resin. The resin is admitted at one end of the mold and is forced by pressure through the mold and preform. Subsequently the resin is cured. This method is limited to resins of low viscosity, such as epoxy. A problem with this process is the formation of surface voids by volatilization of dissolved gases in the resin, partial evaporation of mold releasing agent into the preform, or mechanical entrapment of gas bubbles [38].

A method for forming continuous fiber composites in the shape of cylinders or related objects is filament winding, which involves wrapping continuous fibers from a spool around a form or mandrel. The fibers can be impregnated with a resin before or after winding. Filament winding is used to make pressure tanks. The winding pattern is a part of the composite design. The temperature of the mandrel, the impregnation temperature of the resin, the impregnation time, the tension of the fibers, and the pressure of the fiber winding are processing parameters that need to be controlled. For the case of wet winding of carbon fibers to make epoxy-matrix composites, the optimum temperature of the mandrel is 70–80°C, the optimum resin impregnation temperature is 80–85°C, the sufficient impregnation time is 1–2 sec., the recommended fiber tension is 8.3–16.6 MPa, and the recommended pressure of the winding is 6–8 MPa [39]. For the case of a PEEK-matrix composite, hot air can be used to heat fibers impregnated with the thermoplast, though infrared heating is more effective; a winding speed of up to 0.5 m/sec. has been reported [40].

In most of the composite fabrication methods mentioned above, the impregnation of the fibers with a resin is involved. In the case of a thermoset, the resin is a liquid that has not been polymerized or is partially polymerized.

In the case of a thermoplast, the resin is either the polymer melt or the polymer dissolved in a solvent. After resin application, solid thermoplast results from solidification in the case of melt impregnation, and from evaporation in the case of solution impregnation [41]. Both amorphous and semicrystalline thermoplasts can be melt processed, but only the amorphous resins can normally be dissolved. Because of the high melt viscosities of semicrystalline thermoplasts (e.g., about 370 Pa.sec. for PEEK at 370°C, compared to about 0.39 Pa.sec. for low-viscosity epoxy) due to their long and rigid macromolecular chains, direct melt impregnation of semicrystalline thermoplasts is difficult [35]. Melt impregnation followed by solidification produces a thermoplastic prepreg that is stiff and lacks tack; solution impregnation usually produces prepregs that are drapeable and tacky, although this character changes as solvent evaporation occurs from the solution. The drapeable and tacky character of thermoplastic prepregs made by solution impregnation is comparable to that of thermoset prepregs. Hence, the main problem with resin impregnation occurs for semicrystalline thermoplasts.

Instead of thermoplastic impregnation of fibers by using a melt or a solution of the thermoplast, solid thermoplast in the form of powder, fibers, or slurries can be impregnated [42]. For example, carbon fibers can be immersed in a suspension of a thermoplast powder in an aqueous liquid medium (which contains at least 20 wt.% of an organic liquid) to impregnate the thermoplast into the fibers [43].

An alternative to impregnation is the commingling of continuous carbon fibers with continuous thermoplast fibers. This commingling can be on a fabric level, where yarns of different materials are woven together (coweaving); it can be on a yarn level, where yarns of different materials are twisted together; or it can be on a fiber level, where fibers of different materials are intimately mixed within a unidirectional fiber bundle [44]. During processing, such as compression molding or filament winding, the thermoplast fibers melt, wet the carbon fibers and fuse to form the matrix [45]. However, there is a preferred orientation in the thermoplast fibers, due to the spinning process used in their production, and this may be a problem. Furthermore, the thermoplast fibers have a tendency to form drops during heating [46]. In addition, the availability of high-temperature thermoplast fibers is limited. PEEK is most commonly used for commingling with carbon fibers, but processing must be carried out at a sufficiently high temperature to destroy the previous thermal history of the PEEK matrix [47]. Fiber–matrix adhesion in a commingled system depends on the molding temperature, residence time at the melt temperature, and the cooling rate. This is probably due to several complex mechanisms such as matrix adsorption on the fiber surface, matrix degradation leading to chemical bonding, and interfacial crystallization [47]. On the other hand, prepregs made from commingled fibers are flexible and drapeable [35], and the use of three-dimensional braiding allows net structural shape formation and enhances the damage tolerance due to the lack of delamination [44]. To prevent the ends

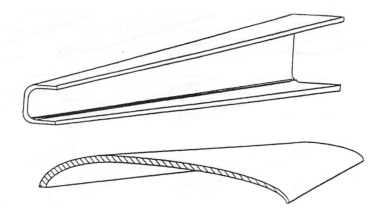

Figure 6.2 Examples of singly curved and long components that can be fabricated by die-less forming. From Ref. 49.

of the braided preform from unbraiding, the thermoplast fibers are melted with a soldering gun before cutting, or alternately, the ends of the braided preform are wrapped with a polyimide tape and cut through the tape. The fiber commingling makes possible a uniform polymer distribution even when the three-dimensional preform is very large, although the heating time during consolidation needs to be longer for dense three-dimensional commingled fiber network braids than for unidirectional prepregs [44].

The shaping of thermoplast-matrix composite laminates can be performed by thermoforming in the form of matched-die forming [48] or die-less forming [49]. However, in addition to shaping, deformations in the form of transverse fiber flow (shear flow perpendicular to the fiber axis) and interply slip commonly occur, while intraply slip is less prevalent [48]. Die-less forming uses an adjustable array of universal, computer-controlled rollers to form an initially flat composite material into a long, singly curved part having one arbitrary cross-sectional shape at one end and another arbitrary shape at the other end, as illustrated in Figure 6.2. Heating and bending of the material are strictly local processes, occurring only within a small active forming zone at any one instant. The initially flat workpiece is translated back and forth along its length in a number of passes. On successive passes, successive portions of the transverse extent of the workpiece pass through the active forming region, as illustrated in Figure 6.3 [49]. Induction heating is used to provide the local heating in die-less forming, because it enables rapid, noncontact, localized and uniform through-thickness heating [50].

The schedule for variation of the temperature and pressure during curing and consolidation of prepregs to form a thermoset-matrix composite must be carefully controlled. Curing refers to the polymerization and cross-linking

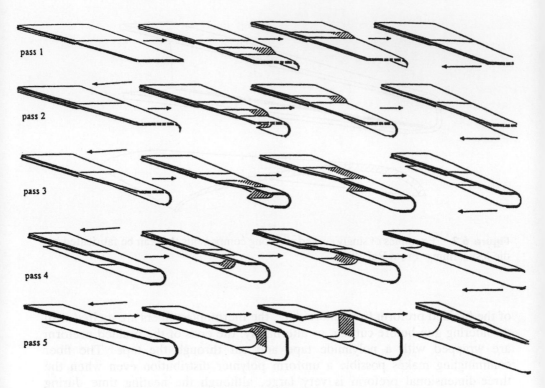

Figure 6.3 Forming sequence in die-less forming of long and tapered components. Hatched areas are active forming regions. (Highly schematic and with an unrealistically large transverse increment per pass.) From Ref. 49.

reactions that occur upon heating and lead to the polymer, whereas consolidation refers to the application of pressure to obtain proper fiber–matrix bonding, low void content, and the final shape of the part. Curing and consolidation are usually performed together as one process. For example, the curing and consolidation of a polyimide-matrix composite involves first heating without pressure at 220°C for 120 min., when melting and imidization occur, and then raising the temperature to 315°C, when the resin initially exhibits melt–flow behavior then solidification. Pressure is applied at the beginning of the 315°C heating stage [51].

The curing of a thermoset-matrix composite requires heat, which is usually obtained by resistance heating, though microwave heating is also possible [52]. An attraction of microwave heating is an increase in the amount of chemical interaction between the carbon fiber surface and the epoxy resin and amine components of the matrix [53].

For the thermoplast-matrix composites, increasing the cooling rate after lamination decreases the crystallinity of the polymer matrix. For cooling rates from 1°F/min. to 1 000°F/min., a PEEK matrix in the presence of carbon fibers has a recrystallinity ranging from 45 to 30 wt.%. Because the fibers act as nucleation sites for polymer crystallization when the polymer melt is sheared, the presence of fibers enhances the polymer crystallinity to a level above that of the neat polymer [54,55]. A greater crystallinity is associated with a higher level of fiber–matrix interaction [56]. The crystallinity can be increased by annealing at up to 310°C; the presence of carbon fibers accelerates the annealing effect [57].

Because of the high processing temperatures (up to 400°C) of high-temperature thermoplasts, traditional tooling materials are not very suitable. Instead of metal tooling materials, carbon fiber polyimide-matrix composites have successfully been used to fabricate parts from prepregs based on polyimide, PEEK, biomaleimides, etc. [58]. The advantages of the composite tooling lies in its low thermal expansion coefficient as well as its thermal stability.

Instead of using heat, thermoplast-matrix composites have been made by aqueous electrocopolymerization onto carbon fibers. An electropolymerization of 3-carboxyphenyl maleimide and styrene onto carbon fibers used dilute sulfuric acid and an aqueous solution containing monomers [59].

Short-fiber composites are formed by mill mixing the short fibers and the polymer powder or by dissolving the polymer in a solvent and forming a carbon fiber paste. The paste method causes less fiber breakage than the mill mixing method [60].

Carbon filaments (made from carbonaceous gases) are typically 0.1 μm in diameter, whereas carbon fibers are typically 10 μm in diameter. Thus, carbon filaments tend to cling together much more than carbon fibers. Consequently, the dispersion of carbon filaments is more difficult than that of carbon fibers. For the case of a thermoset matrix, the carbon filaments may be impregnated with the liquid resin, which subsequently sets; this process is not difficult. However, for the case of a thermoplast matrix, the thermoplast is usually in the form of particles, and the mixing of the thermoplast particles with the fine carbon filaments to achieve good filament dispersion may be difficult. A method to achieve filament dispersion in a thermoplast involves (1) dispersing the filaments in an alcohol aqueous solution with the help of a trace amount of a dispersant, (2) mixing the slurry with the thermoplast powder at room temperature by using a rotary blade blender to adjust the concentration of alcohol so that the thermoplast particles are suspended in the aqueous solution, (3) draining the solution, (4) drying, and (5) hot pressing above the T_g of the thermoplast. The mixing in step (2) causes very little filament breakage, so the aspect ratio of the filaments after mixing remains high (at least 1 000). After Step 4, a dry and uniform mixture of the thermoplast particles and the filaments is obtained in the form of clusters of size 0.1–0.5 mm. These clusters can be conveniently handled. They can be put into a die for hot pressing, (Step 5), which results in a thermoplast-matrix composite [61].

Properties

Carbon fiber polymer-matrix composites have the following attractive properties [62]:

- low density (40% lower than aluminum)
- high strength (as strong as high-strength steels)
- high stiffness (stiffer than titanium, yet much lower in density)
- good fatigue resistance (a virtually unlimited life under fatigue loading)
- good creep resistance
- low friction coefficient and good wear resistance (a 40 wt.% short carbon fiber nylon-matrix composite has a friction coefficient nearly as low as Teflon and unlubricated wear properties approaching those of lubricated steel)
- toughness and damage tolerance (can be designed by using laminate orientation to be tougher and much more damage tolerant than metals)
- chemical resistance (chemical resistance controlled by the polymer matrix)
- corrosion resistance (impervious to corrosion)
- dimensional stability (can be designed for zero coefficient of thermal expansion)
- vibration damping ability (excellent structural damping when compared with metals)
- low electrical resistivity
- high electromagnetic interference (EMI) shielding effectiveness
- high thermal conductivity

The mechanical properties and density of carbon fiber epoxy-matrix composites compared to other materials are shown in Table 6.6 [62]. The fatigue behavior (S–N curve) of a carbon fiber epoxy-matrix composite is shown in Figure 6.4 [62]. It is superior to that of Kevlar fiber epoxy-matrix composite, boron fiber epoxy-matrix composite, boron fiber aluminum-matrix composite, glass fiber epoxy-matrix composites, and aluminum (2024-T3) [62]. Furthermore, it is superior to that of steel [37].

The mechanical properties of unidirectional carbon fiber (Hercules, Magnamite AS-4) epoxy-matrix composites are shown in Table 6.7 for a fiber content of 62 vol.% [63]. The tensile strength and modulus (along the fiber direction) are very close to those calculated from the fiber strength (3 795 MPa) and fiber modulus (235 GPa) using the rule of mixtures.

The mechanical properties of unidirectional carbon fiber composites are obviously anisotropic. The tensile strength and modulus are much higher in the longitudinal (0°) direction than the transverse (90°) direction, though the ultimate strain is higher in the transverse direction than the longitudinal direction. The compressive strength and modulus are much higher when the compressive stress is perpendicular to the fiber layers (0°) than when the compressive stress is parallel to the fibers (90°); this is because stress parallel to

Table 6.6 Mechanical properties and density of unidirectional carbon fiber epoxy-matrix composite compared to other unidirectional epoxy-matrix composite materials and metals. From Ref. 62.

	Strength[a] (MPa)		Tensile modulus (GPa)	Density (g/cm³)
Material	Tension	Compression		
Epoxy/carbon fibers AS-4	1 482	1 227	145	1.55
Epoxy/carbon fibers HMS	1 276	1 020	207	1.63
Epoxy/S-2 glass fibers	1 751	496	59	1.99
Epoxy/E-glass fibers	1 103	490	52	1.99
Epoxy/Aramid Kevlar 49	1 310	290	83	1.39
Aluminum (7075-T6)	572	–	69	2.76
Titanium (6Al-4V)	1 103	–	114	4.43
Steel (4130)	1 241–1 379	–	207	8.01

[a]Unidirectional.

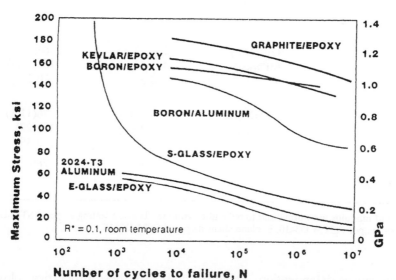

Figure 6.4 *S–N* curve to show the fatigue behavior of unidirectional composites and aluminum. From Ref. 62.

Table 6.7 Mechanical properties of unidirectional carbon fiber (62 vol.% AS-4) epoxy-matrix composites. From Ref. 63.

Tensile strength	2 353 MPa
Tensile modulus	145 GPa
Flexural strength	1 794 MPa
Flexural modulus	131 GPa
Short-beam shear strength	124 MPa

Table 6.8 Room temperature mechanical properties of carbon fiber epoxy-matrix composites with P-75 fibers and 934 epoxy. From Ref. 64.

Properties/test method	Unit	Unidirectional	Quasi-isotropic
0° Tension[a]			
Modulus	GPa (Msi)	310(45)	103(14.9)
Strength	MPa (ksi)	999(145)	246(35.8)
Ult. strain	%	0.31	0.24
90° Tension[a]			
Modulus	GPa (Msi)	7.6(1.1)	108(15.7)
Strength	MPa(ksi)	21.3(3.1)	351(51.0)
Ult. strain	%	0.3	0.33
0° Compression[a]			
Modulus	GPa(Msi)	229(33.3)	66(9.6)
Strength	MPa(ksi)	421(61.2)	183(26.7)
Ult. strain	%	0.32	0.5
90° Compression[a]			
Modulus	GPa(Msi)	7.6(1.1)	76(11.1)
Strength	MPa(ksi)	129(18.8)	186(27)
Ult. strain	%	–	0.55
Short-beam shear strength	MPa(ksi)	63(9.1)	28(4.1)

[a]Strength and modulus are normalized to 60% fiber volume. Tension testing as per ASTM D3039, compression as per ASTM D3410, in-plane shear as per ASTM D3518.

the fibers causes delamination. Table 6.8 [64] lists these properties, plus the short-beam shear strength, for a unidirectional composite with Amoco's Thornel P-75 as the fibers and epoxy 934 as the matrix.

Quasi-isotropic continuous-fiber composites can be obtained by laying the fibers in different layers in different directions. Table 6.8 [64] also lists the mechanical properties of a composite with P-75 fibers having the stacking

Table 6.9 Properties of unidirectional carbon fiber (AS4-3K) thermoplast-matrix composites, with each composite having 10 plies of prepreg tape. From Ref. 65.

Matrix	PEEK	PPS
Panel thickness per ply (mm)	0.13	0.14
Specific gravity	1.56	1.53
Fiber vol.%	60.0	56.2
Void vol.%	1.9	1.4
Flexural properties		
Strength (MPa)	1 687	1 078
Modulus (GPa)	108	93.8
Transverse tensile strength[a] (MPa)	91.0	15.2
H_2O absorption (wt.%)	0.15	0.20
Coefficient of thermal expansion (10^{-6}/°C)		
-157 to 21°C	0.18	-1.1
21 to 121°C	0.49	0.18
Outgassing		
Percent total mass loss	0.0348	0.0291
Percent collected volatile condensable materials	0.0054	0.0046

[a]ASTM D3039-74.

sequence $(0°, 30°, 60°, 90°, 120°, 150°)s$ and with epoxy 934 as the matrix. The properties are indeed quasi-isotropic. The in-plane coefficient of thermal expansion of this composite is -0.16×10^{-6}/°C [64].

The properties of unidirectional carbon fiber thermoplast-matrix (PEEK, PPS) composites are listed in Table 6.9 [65]. Comparison of Tables 6.9 and 6.7 show higher flexural strength and modulus for an epoxy-matrix composite than for a PEEK-matrix composite of similar fiber content and the same fiber type. On the other hand, the residual compressive strength after impact (CAI) of the PEEK-matrix composite is comparable to that of a toughened-epoxy-matrix composite and is superior to those of composites with standard epoxy or standard bismaleimide (BMI) matrices, as shown in Figure 6.5 [66]. The superior impact resistance of thermoplast-matrix composites compared to thermoset-matrix composites makes thermoplast-matrix composites particularly attractive for aircraft applications. It should be mentioned that the mechanical properties of carbon fiber PEEK-matrix composites are influenced by the matrix crystallinity and the cooling rate, as shown in Table 6.10 for unidirectional carbon fibers (Hercules AS-4) [46].

The water absorption of the thermoplast-matrix composites is much lower than that of the epoxy-matrix counterpart (Figure 6.6). The equilibrium moisture levels are 0.15%, 0.20%, and 2.32% for PEEK-, PPS-, and epoxy-matrix composites, respectively [65]. The outgassing amounts are comparable for composites of these three matrices and are all low enough to be acceptable for spacecraft applications.

Table 6.10 Effect of matrix crystallinity and cooling rate on the mechanical properties of unidirectional carbon fiber (AS-4) PEEK-matrix composites. From Ref. 46.

| | | Flexural strength (MPa) | | ILSS | Young's modulus |
Crystallinity	Cooling rate	Longitudinal	Transverse	(MPa)	(GPa)
High	High	2 002	146	101	111.1
High	Low	2 066	114	103	101.6
Low	High	2 162	169	107	106.4
Low	Low	2 426	131	108	128.0

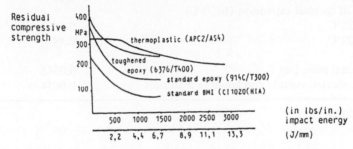

Figure 6.5 Residual compressive strength after impact versus impact energy for carbon fiber composites with various polymer matrices. From Ref. 66. (Reprinted by permission of Kluwer Academic Publishers.)

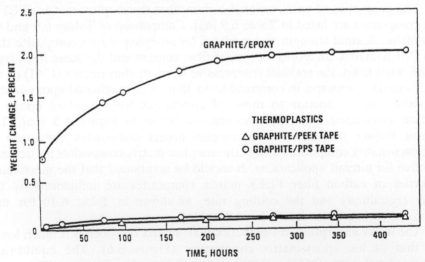

Figure 6.6 Water absorption at 160°F (71°C) for carbon fiber polymer-matrix composites with epoxy and thermoplasts (PEEK and PPS) as matrices. From Ref. 65. (Reprinted by permission of the Society for the Advancement of Material and Process Engineering.)

Table 6.11 Transverse (±45°) tensile properties of carbon fiber PEEK-matrix composites. From Ref. 67

Reinforcement dimensionality	Strength (MPa)	Modulus (GPa)	Fracture strain[a] (%)
1	309	7.72	26
2	255	8.00	14
3	155	8.34	7.0

[a]Measured by crosshead travel.

Table 6.9 shows a higher transverse tensile strength for the PEEK-matrix composite than the PPS-matrix composite. This is because of the stronger fiber–matrix adhesion in the former. The stronger fiber–matrix adhesion for PEEK is due to the higher crystallinity in PEEK than PPS [65].

The transverse (±45°) tensile properties of PEEK composites containing unidirectional carbon fiber tape, two-dimensional fabric, and three-dimensional fabric are shown in Table 6.11 [67]. Both the strength and fracture strain decrease with increasing dimensionality, whereas the modulus does not vary much with dimensionality.

The use of carbon fiber polymer-matrix composites for primary structural applications requires high compressive strength after impact damage (CAI) and high compressive strength at elevated temperatures after exposure to wet environments (CHW). These properties depend on the toughness of the matrix resin. Methods to improve the matrix toughness include the following [68]:

- using an epoxy resin of high elongation
- blending an elastomer with the epoxy resin
- blending a thermoplastic with the epoxy resin
- introducing a discrete interleaf layer between prepreg layers in the laminate
- using a tough thermoplastic matrix

The damping ability of a fibrous composite improves as the fiber–matrix bonding weakens. Figure 6.7 [65] shows that the damping ability (loss factor) increases in the order: epoxy, PEEK, PPS.

The wear resistance and friction coefficient of carbon fiber polymer-matrix composites depend on the sliding direction with respect to the fiber direction. For unidirectional fiber composites, the fibers can be in the plane of sliding and parallel to the direction of sliding (termed longitudinal); they can be in the plane of sliding and perpendicular to the direction of sliding (termed transverse); they can stand normal to the plane of sliding (termed normal). Table 6.12 shows the wear factors and friction coefficients of high-modulus

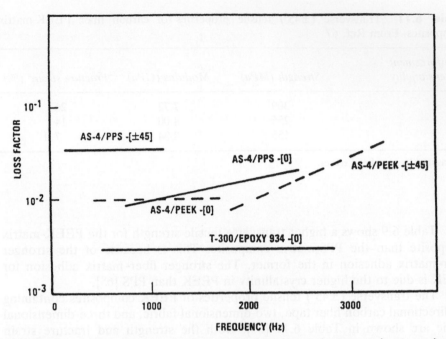

Figure 6.7 Loss factor versus frequency of carbon fiber polymer-matrix composites with epoxy and thermoplasts (PEEK and PPS) as matrices. From Ref. 65. (Reprinted by permission of the Society for the Advancement of Material and Process Engineering.)

unidirectional carbon fiber epoxy-matrix composites sliding against stainless steel for these three directions. The wear resistance is worst for the normal direction. The wear resistance is best and the friction coefficient is least in the transverse direction [69]. However, for unidirectional carbon fiber PEEK-matrix composites sliding against a 180 grit SiC abrasive paper under a contact pressure of 2.2 MPa and at a sliding velocity of 0.3 m/min., the wear rate is higher in the transverse direction than in the other two directions, while the

Table 6.12 Wear factor and friction coefficient of unidirectional carbon fiber epoxy-matrix composites sliding against stainless steel for 48 h. From Ref. 69.

Condition	Orientation	Wear factor $(10^{-11}\ g/Nm)$	Friction coefficient
Load = 49 N	Longitudinal	4.79	0.30
Speed = 0.5 m/s	Transverse	3.54	0.22
	Normal	5.36	0.29
Load = 19.6 N	Longitudinal	9.48	0.17
Speed = 2.5 m/s	Transverse	4.07	0.13
	Normal	22.56	0.18

friction coefficient is essentially independent of the sliding direction [70]. In general, carbon fibers improve the wear resistance of polymers; this is because of the strengthening due to the fibers and, in the case of high-modulus carbon fiber composites, because the composite generates a surface wear film, which reduces friction and wear owing to the self-lubricating nature of the graphite debris. Both unreinforced PEEK and a carbon fiber PEEK-matrix composite display a relative minimum in wear against mild steel at a certain level of counterface roughness, but the minimum for the composite occurs at a greater counterface roughness than the minimum for the unreinforced PEEK [71].

The plastic memory phenomenon is the tendency of a thermoplastic material that has been deformed above T_g to return to its original shape upon reheating above T_g. The origin of this phenomenon probably is related to the fact that the potential energy, which the polymer chains contain in the deformed shape (in the glassy solid phase), is released when the specimen is reheated into the rubbery phase. The polymer chains probably dissipate their stored energy by relative movement, eventually reaching their lowest energy configuration corresponding to the specimen's originally molded shape. This phenomenon is exhibited not only by the neat thermoplast, but also by thermoplasts, PEEK and polybutylene teraphthalate (PBT), reinforced by continuous or short carbon fibers, though the use of neat thermoplast laminae in addition to the carbon fiber prepreg laminae markedly improves the shape recovery. The plastic memory phenomenon is potentially useful for self-deploying of space structures such as antenna reflectors [72].

The high thermal conductivity of carbon fibers, especially the high modulus pitch-based fibers (Amoco's Thornel P-100 and P-120, with fiber thermal conductivity at 300 K of 300 and 520 W/m/K, respectively) and the vapor grown carbon fibers (with fiber thermal conductivity at 300 K of 1 380 W/m/K), makes these fibers highly effective for increasing the thermal conductivity of polymers. Table 6.13 shows the 300 K thermal conductivity and the thermal conductivity/density ratio of various unidirectional carbon fiber polymer-matrix composites, together with the corresponding values of metals. The highest thermal conductivity at 245 W/m/K is associated with the P-120 fibers. This conductivity value is higher than that of aluminum, though lower than that of copper. The ratio of the thermal conductivity to the density is higher than that of aluminum for composites with P-120 or P-100 fibers in the amount of 45 vol.%. The combination of high thermal conductivity, low density, and good corrosion resistance makes these composites valuable for aerospace structures, electronic packaging, and many other applications [73]. One disadvantage of using P-100, P-120, or other graphitic carbon fibers of high thermal conductivity is that they cause the polymer-matrix composite to be so low in thermal expansion that CTE mismatch occurs between the composite and its neighbor (e.g., a printed circuit board). In order to alleviate this problem, metal-coated carbon fibers are used [74].

If high-strength carbon fibers rather than high-modulus carbon fibers are used, the thermal conductivity of the composites are lower. Figures 6.8 and 6.9

Table 6.13 Thermal conductivities (κ) of the composites compared with those of copper and aluminum. From Ref. 73.

Sample	Fibers	Matrix	Fiber (%)	Density (kg/m^3)	κ_{300K} ($W/m/K$)	$\kappa_{300K}/density$ ($W.m^2/kg/K$)
#3	P-75	Polystyrene	35	1 340	59.8	0.045
#4	P-75	Polystyrene	30	1 290	30.1	0.023
#5	P-100	Polystyrene	30	1 340	95.4	0.072
#6	P-100	Polyester	15	1 370	60.0	0.044
#8	P-75	Polyester	29	1 450	45.8	0.032
#9	P-75	Polyester	45	1 580	64.5	0.042
#10	P-100	Polyester	45	1 640	140.4	0.088
#11	P-120	Polyester	45	1 660	245.0	0.148
Pure copper				8 960	450.0	0.050
Pure aluminum				2 700	200.0	0.074
Stainless steel				7 860	15.0	0.002

show the thermal conductivities of unidirectional high-strength carbon fiber (50 vol.%) epoxy-matrix composites in directions parallel and perpendicular to the fibers for temperatures from 300 K down to 4.2 K. Also shown in Figure 6.8 are the corresponding data for unidirectional epoxy-matrix composites with 50 vol.% glass fibers, 50 vol.% SiC fibers, and 50 vol.% Al_2O_3 fibers. In the direction parallel to the fibers (Figure 6.8), the carbon fiber composite has thermal conductivity that is higher than the other three composites at temperatures above 40 K, but has thermal conductivity that is lower than glass and SiC fiber composites and comparable to the Al_2O_3 fiber composite at temperatures below 40 K; in particular, at 300 K the thermal conductivity of the carbon fiber composite is about four times as much as that of the glass fiber composite. In the direction perpendicular to the fibers (Figure 6.9), the thermal conductivities of all four fiber composites are similar for the whole temperature range from 300 K down to 4.2 K, so the thermal conductivity in this direction is dominated by the epoxy matrix. Comparison of Figures 6.8 and 6.9 shows that the thermal conductivity in the direction perpendicular to the fibers is lower than that in the direction parallel to the fibers for all four types of fiber composite [75].

Short carbon fibers are commonly used as an electrically conductive filler for polymers used for electromagnetic interference shielding, antistatic, and other electronic applications. Due to their small diameter, short carbon filaments tend to have a larger aspect ratio than short carbon fibers. As a result, short carbon filaments tend to be more effective at providing electrically conductive polymer-matrix composites than comparably short carbon fibers of similar electrical resistivity. This means that a lower volume fraction of carbon

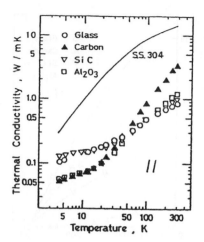

Figure 6.8 Thermal conductivity of 304 stainless steel and various fiber epoxy-matrix composites (50 vol.% fibers) in the fiber direction from 300 K down to 4.2 K. From Ref. 75. (By permission of Plenum Publishing Corp.)

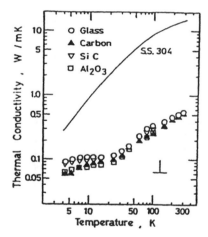

Figure 6.9 Thermal conductivity of 304 stainless steel and various fiber epoxy-matrix composites (50 vol.% fibers) in the transverse direction from 300 K down to 4.2 K. From Ref. 75. (By permission of Plenum Publishing Corp.)

filaments is required to attain the same level of electrical resistivity of the composite compared to carbon fibers [61].

Figures 6.10 and 6.11 show the thermal contraction of unidirectional high-strength carbon fiber (50 vol.%) epoxy-matrix composites in directions parallel and perpendicular to the fibers for temperatures from 300 K down to 77 K. Also shown in these figures are the corresponding data for composites

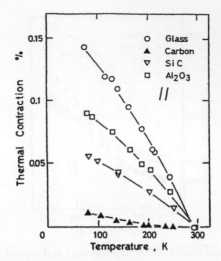

Figure 6.10 Thermal contraction of various fiber epoxy-matrix composites (50 vol.% fibers) in the fiber direction from 300 K down to 77 K. From Ref. 75. (By permission of Plenum Publishing Corp.)

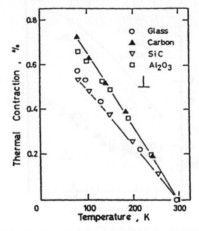

Figure 6.11 Thermal contraction of various fiber epoxy-matrix composites (50 vol.% fibers) in the transverse direction from 300 K down to 77 K. From Ref. 75. (By permission of Plenum Publishing Corp.)

with glass, SiC and Al_2O_3 fibers instead of carbon fibers. In the direction parallel to the fibers (Figure 6.10), the thermal contraction is lowest for the carbon fiber composite, which has a thermal contraction of 0.015% at 77 K. In the direction perpendicular to the fibers (Figure 6.11), the thermal contraction is much larger than in the direction parallel to the fibers and is similar for all four fiber types [75].

The cryogenic properties of carbon fiber composites are relevant to the application as cryogenic structural support members. This application is made possible by the fact that the strength and modulus of the composites increase with decreasing temperature [75,76].

The low electrical resistivity and high aspect ratio of short carbon fibers make them very effective as a filler for making electrically conducting polymer-matrix composites, which are useful for electromagnetic interference (EMI) shielding [77] and, in the case of pressure-sensitive conductive rubber composites, for touch control switches and strain sensors [78]. For electrically conducting polymer-matrix composites with a discontinuous filler (which allows injection molding and related composite fabrication methods), short carbon fibers compete with carbon black as the filler. Due to the high aspect ratio of short carbon fibers compared to carbon black, the required critical filler volume fraction is lower for short carbon fibers than carbon black. However, for the case of rubber as the matrix, due to the stronger filler–matrix interaction for carbon black as the filler, the mechanical properties are superior for composites containing carbon black [79]. The EMI shielding effectiveness and electrical conductivity of carbon fiber polymer-matrix composites increase with increasing fiber content and decrease with increasing frequency [77,80]. The frequency dependence of the AC conductivity (with the electric field parallel to the fibers) is due to the reactance increase associated with the magnetic flux change from neighboring fibers [80]. For pressure-sensitive conductive rubber composites, the electrical conductivity increases but the pressure sensitivity decreases when the fibers are oriented along the longitudinal direction with respect to the electric field, whereas the electrical conductivity decreases but the pressure sensitivity increases when the fibers are transversely oriented [78]. The electrical conductivity and EMI shielding effectiveness can be enhanced by using nickel-coated carbon fibers [81]. Nickel is more commonly used than copper for the coating because of the superior oxidation resistance of nickel.

Degradation

The degradation of carbon fiber polymer-matrix composites under hygrothermal environments causes deterioration of the mechanical properties. This is mostly due to a decrease in the fiber–matrix adhesion. For unidirectional carbon fiber epoxy-matrix composites, the mechanical property deterioration is in terms of the transverse properties (transverse modulus, ILSS, etc.) rather than the longitudinal properties [82]. Table 6.14 [83] shows the relative ILSS (i.e., the ratio of the wet ILSS to the dry ILSS) as a function of time of immersion in water at 50°C for untreated carbon fiber and styrene/air plasma-treated fiber in an epoxy matrix. The degradation is particularly fast during the first hour of water immersion. Beyond 1 h., the degradation continues, but more slowly. The nearly stabilized state of degradation corresponds to a relative ILSS of 32% for the untreated fiber and 67% for the plasma-treated fiber [83]. The deterioration becomes more severe as the

Table 6.14 Relative ILSS after immersion of epoxy-matrix composite in water at 50°C. From Ref. 83.

Fiber	Immersion time (h.)	Relative ILSS (%)
Untreated	0	–
	40	51.8
	60	42.5
	90	34.0
	131	31.9
Styrene/air plasma treated	0	–
	28	81.5
	69.5	67.9
	130.5	66.7

temperature increases. The fatigue resistance is particularly sensitive to moisture, as the fatigue life is significantly lowered even for less than 2% moisture in the epoxy-matrix composite [84].

In contrast to epoxy matrix, PEEK and PPS matrices yield composites that retain their mechanical properties under hot/wet conditions. In particular, carbon fiber PEEK-matrix composites retain 95% of the flexural strength after 1 400 h. at 80°C and 75% RH; the saturated water absorption is 0.136 wt.%; the diffusion coefficient of water under this hot/wet condition is 8.43×10^9 cm^2/sec [85].

Moisture absorption in unidirectional carbon fiber epoxy-matrix composites causes swelling of up to 0.70% in the in-plane transverse direction, up to 1.91% in the out-of-plane direction, and up to 2.61% in volume expansion. The dimensional change is undesirable for applications such as the supporting structures of spacecraft antennas and telescopes [86].

Carbon fiber epoxy-matrix composite parts may be joined by using epoxy film adhesives. Such joints degrade due to water absorption by the adhesives. The degradation corresponds to a retention of 60% of the dry strength after exposure to 50°C and 96% RH for 3 years [87].

Thermal aging can degrade the mechanical properties of carbon fiber polymer-matrix composites due to the oxidation of the carbon fibers and of the matrix. The extent of degradation is less when the fiber–matrix bonding is stronger, as air could permeate a weak fiber–matrix interface. Among composites with high-temperature polymer matrices, the key to thermal stability is the selection of fibers such that the fiber–matrix bonding is strong; it is less important to select fibers which are by themselves more oxidation resistant. Indeed the degradation in shear strength of carbon fiber polyimide-matrix composites (PMR-15) after aging in air at 316°C for $\geq$ 1 188 h. is not due to the weight loss but rather to changes in fiber–matrix interaction, perhaps caused by small changes in resin composition or by chemical changes at the fiber–matrix interface [88]. Moreover, the activation energy of oxidation of the composite is greater than those of the fibers and the matrix, indicating a synergistic effect from the fiber–matrix combination [89]. By using *N*-phenylnadimide-modified PMR-15 polyimide and Celion 6000 carbon fibers, composites that can be used for 100 h. under continuous heating at 371°C in air have been reported. The superior thermal stability of the modified PMR-15 composite compared to the counterpart without modification is attributed to the higher residual stress in the latter due to the greater cross-link density [90]. For PEEK and BMI as matrices, it has been reported that uni-directional carbon fiber composites have higher percentage retentions of ILSS and impact strength than the multidirectional counterpart after thermal aging in air at 190°C for up to 1 000 h. [91].

Thermal spikes from 20°C to 150°C (as can be produced in the outer skins of supersonic aircraft) decrease considerably the fatigue life of carbon fiber epoxy-matrix composites, even when no damage is observed by optical microscopy [92]. Impact at ultrahigh strain rates of about 10^7/sec., as obtained by short pulsed laser induced shock waves, cause spall damage in carbon fiber epoxy-matrix composites at a threshold pressure of 1.5 kbar, compared to 21 kbar for aluminum and 54 kbar for iron [93].

Carbon fiber polymer-matrix composites give rise to galvanic corrosion when they are in contact with metals (other than platinum, gold, and titanium) which get corroded because the carbon fibers serve as the cathode. When the composite is in seawater, aragonite crystal (a form of calcium carbonate) grows on the composite, causing blistering. Due to the common use of metals as substructures and fasteners in systems in the ocean, this problem is frequently encountered [94].

High-energy radiation encountered by composites used for space satellites in geosynchronous orbit can cause degradation due to the molecular scissions in the polymer matrix. This problem can be alleviated by the incorporation of radiation-resistant groups into the polymer [95].

Joining and Repair

The joining and repair of composite parts are closely related, as joining is usually involved in repair. Joining methods differ greatly between thermoset- and thermoplast-matrix composites.

Thermoset-matrix composites are commonly joined by using an adhesive film, such as epoxy, though the joint strength is low. Mechanical fastening is sometimes used, but it is difficult because the drilling of holes in the brittle composite causes weakness and even damage, and the metals used as screws cause thermal expansion mismatches and induce galvanic corrosion due to the cathodic nature of carbon. Thus, the joining of thermoset-matrix composites is an area which needs further work.

Thermoplast-matrix composites are commonly joined by fusion welding, which means heating to cause flow between the thermoplast parts. As the temperatures involved are much lower than those required for the fusion welding of metals, the techniques of thermoplast welding differ from those of metal welding. The techniques of thermoplast fusion welding include electrical resistance heating, focused infrared heating, vibration (ultrasonic) welding, the heated press technique and thermoplastic interlayer bonding. In general, the joining techniques utilize heat to melt either the matrix resin or an unreinforced thermoplastic film placed between the parts to be joined. The interlayer (interleaf) can be a 100 μm thick film of amorphous PEEK [96] or amorphous polyetherimide (PEI) [97], in the case of joining PEEK-matrix composite parts. Lap shear joint strengths in excess of 50 MPa have been achieved; most successful was the heated press technique [96]. The fusion welding methods are rapid (5 min.) compared to the use of an epoxy film adhesive (1 h.) [97].

The repair of damaged carbon fiber thermoplast-matrix composites can be achieved by hot pressing at the usual molding temperature for the thermoplast. For PEEK as the thermoplast matrix, delamination-type fracture such as that incurred at low incident energies can be fully repaired without any loss in mechanical integrity. However, in the case of extensive fiber fracture (as incurred at energies of 12 J or more), the success of the repair is not total, though the residual compressive and flexural strengths of the repaired specimen are still considerably greater than those of the as-impacted specimen [98].

Inspection

The inspection of fiber composite parts is mainly for observing the fiber arrangement and defects, as such structural features strongly affect the properties of the composites. Defects include the following [99]:

- matrix cracks (voids, porosities)
- fiber cracks
- interface cracks (debonding)
- delamination (splitting between laminae and a laminate)
- inclusions (foreign bodies in the composite)

Techniques for inspection include metallography (microscopy of polished surfaces), transmission microscopy of thin sections, low-voltage radiography with soft X-rays [99], infrared thermal imaging and the C-scan technique [98]. The infrared technique makes use of the thermally conductive nature of carbon fibers, but it is mostly limited to continuous fiber laminates [100].

Applications

Carbon fiber polymer-matrix composites are predominantly used for the aerospace industry, but the decreasing price of carbon fibers is widening the applications of these composites to include the automobile, marine, sports, biomedical, construction, and other industries.

One area of aerospace applications is space vehicles. The United States Space Shuttle uses carbon fiber epoxy-matrix composites for its payload bay door and remote manipulator arm [101]; its solid rocket motor cases also use epoxy-matrix composites; its booster tail and fins use polyimide-matrix composites. Satellite structures [102–104] and solar panels also use carbon fiber polymer-matrix composites. Most space applications utilize standard aerospace grade carbon fibers (tensile strength 3 550 MPa, tensile modulus 235 GPa) combined with a 177°C cure multifunctional epoxy resin matrix. Filament wound rocket motor cases employ a 121°C cure, modified bis-A-epoxy as the resin matrix. Stiffness requirements of some satellite applications dictate the use of high-modulus carbon fibers (350 GPa) [105]. Thermoplast matrices such as PEEK [106] and PES [107] are gaining attention for space applications.

A second area is military aircraft. Examples include Gripen, EFA, French Rafale, and U.S. B-2, which use the 177°C cure toughened thermoset matrix resins along with intermediate-modulus (295 GPa), or high-strength (5 590 MPa), intermediate-modulus carbon fibers. The U.S. Advanced Tactical Fighter is planned to use thermoplastics or toughened bismaleimide as a primary structural material. Older military aircraft are being modified with epoxy-matrix composite wings [105].

Helicopters are a third area, both for military and commercial use. For example, the all-composite MBB BK117 helicopter is two-thirds carbon fiber epoxy-matrix composite, one-third aramid fiber and glass fiber epoxy-matrix composite. Both 121°C and 177°C cure epoxy resins are used [105].

A fourth area is concerned with primary and secondary structures in commercial aircraft [108]. Examples of primary structural applications include the Airbus A310/A320 vertical tail fin boxes (121°C cure toughened epoxy), the A320 horizontal tail fin (177°C cure epoxy) [108] and the ATR 72 external wing box (177°C toughened epoxy) [105].

A fifth area of aerospace applications is commercial aircraft engines. Outer and front sections of the engine are subjected to lower temperatures and can utilize an epoxy matrix. For example, the front fan ducts on Rolls-Royce engines and the blocker doors and transcowls on General Electric's CF6-80C2 engines use 177°C cure epoxy. Engine rear section components operate at higher temperatures; this necessitates polyimide matrices such as PMR-15, which is used for thrust reversers and bypass ducts. Thermoplasts such as PEEK are being considered for engine applications [105].

Aluminum is a lightweight metal that competes with carbon fiber polymer-matrix composites for aerospace applications. In addition to their much higher strength and modulus, the carbon fiber composites are produced using much less energy and costly pollution control compared to aluminum [109].

Carbon fiber polymer-matrix composites have started to be used in automobiles mainly for saving weight (i.e., fuel economy). The so-called graphite car employs carbon fiber epoxy-matrix composites for body panels, structural members, bumpers, wheels, drive shaft, engine components, and suspension systems [110,111]. This car is 1 250 lb. (570 kg) lighter than an equivalent vehicle made of steel. It weighs only 2 750 lb. instead of the conventional 4 000 lb. for the average American car [110]. Thermoplastic composites with PEEK and polycarbonate (PC) matrices are finding use as spring elements for car suspension systems [112].

The electrically conductive characteristic of carbon fiber polymer-matrix composites makes them suitable for static dissipation (which requires an electrical resistivity of 10^4–10^6 Ω.cm), functional elements in high-impedance circuits (which require a resistivity of 10^2–10^3 Ω.cm), and shielding from radio frequency interference (which requires a resistivity of 10^0–10^2 Ω.cm). From loadings as low as 10 wt.%, a polymer is made static-dissipating, protecting electronic circuits or avoiding spark generation [113]. In addition, carbon fiber polymer-matrix composites are used for RF components [114]. The protection of aircraft from lightning damage is a related application [115,116]. The electrically conductive characteristic also makes carbon fiber polymer-matrix composites useful as electrodes [117,118].

The high thermal conductivity and low thermal expansion of continuous carbon fiber polymer-matrix composites make them suitable for heat sinks in electronics [119]. Since a heat sink is in contact with a ceramic chip carrier or a printed circuit board, a low thermal expansion is preferred. The low density of the composites (compared to copper) makes them even more attractive for aerospace electronics [120].

The X-ray transparency of carbon fibers makes carbon fiber polymer-matrix composites useful for passing small-impulse electric currents to monitor a patient's vital signs while he is being X-rayed [121].

Thermoplasts filled with short or continuous carbon fibers are useful as bone plates for fracture fixation. Metal bone plates suffer from metallic ion leaching, which may cause adverse local tissue reactions and even local tumor formation, and from stress protection atrophy. Polylactic acid (PLA) is an absorbable thermoplast used for this application, but its mechanical properties are not sufficient for long bone fixation, so continuous carbon fibers are added to produce a semiabsorbable composite [122]. Polymers, such as PEEK, which are not absorbable are also used for this application [123].

Due to the concern about the loss of bone around stiff metallic femoral stems, more flexible carbon fiber polymer-matrix composites are being considered for use in hip replacement prostheses [124].

Continuous carbon fiber polymer-matrix composites are replacing steel for reinforcing concrete structures, because the composites are lightweight, available in continuous and long lengths, and do not rust. The lightweight characteristic makes them convenient to install [125,126].

Continuous carbon fiber polymer-matrix composites are used as acoustic diaphragms in speakers and microphones because of their low weight, high elasticity, fast sound transmission velocity, and excellent rigidity. These diaphragms exhibit less deformation due to an external force, a small sound distortion, wide sound reproduction range, distinct sound quality, and are suitable for digital audio applications [127].

Short carbon fibers, together with graphite powder, in a polyimide matrix provide an abrasion-resistant material that is useful for bearings [128].

Short carbon fibers in a polyurethane resin or its precursor provide a sealing compound with a high tensile strength for use in filling and sealing a gap between two parts [129].

References

1. W.-T. Whang and W.-L. Liu, *SAMPE Q.* **22**(1), 3–9 (1990).
2. D.C. Sherman, C.-Y. Chen, and J.L. Cercena, in *Proc. Int. SAMPE Symp. and Exhib.*, *33, Materials: Pathway to the Future*, edited by G. Carrillo, E.D. Newell, W.D. Brown, and P. Phelan, 1988, pp. 134–145.
3. D.R. Askeland, *The Science and Engineering of Materials*, 2nd ed., PWS-Kent, 1989, pp. 538–539.
4. S.D. Mills, D.M. Lee, A.Y. Lou, D.F. Register, and M.L. Stone, in *Proc. 20th Int. SAMPE Tech. Conf.*, 1988, pp. 263–270.
5. S. Wang and A. Garton, *Polym. Mater. Sci. Eng.* **62**, 900–902 (1990).
6. T.E. Twardowski and P.H. Geil, *J. Appl. Polym. Sci.* **42**(6), 1721–1726 (1991).
7. H.G. Recker, *SAMPE J.* **26**(2), 73–78 (1990).
8. W.D. Bascom, *Polym. Mater. Sci. Eng.* **63**, 676–680 (1990).
9. R.B. Gosnell, in *Proc. Int. SAMPE Symp. and Exhib.*, *33, Materials: Pathway to the Future*, edited by G. Carrillo, E.D. Newell, W.D. Brown, and P. Phelan, 1988, pp. 746–753.
10. R.D. Vannucci, D. Malarik, D. Papadopoulos, and J. Waters, in *Proc. Int. SAMPE Tech. Conf.*, *22, Advanced Materials: Looking Ahead to the 21st Century*, edited by L.D. Michelove, R.P. Caruso, P. Adams, and W.H. Fossey, Jr., 1990, pp. 175–185.
11. Z. Qusen, L. Yuhua, C. Zhenghua, and S. Dongsheng, MD, 5(*Adv. Compos. Process. Technol.*), 27–32 (1988).
12. R.J. Morgan, R. Jurek, D.E. Larive, C.M. Tung, and T. Donnellan, *Polym. Mater. Sci. Eng.* **63**, 681–685 (1990).
13. M.S.M. Alger and R.W. Dyson, *Engineering Polymers*, edited by R.W. Dyson, Blackie, Glasgow, 1990, pp. 1–28.
14. R.E. Allred and L.A. Harrah, in *Proc. Int. SAMPE Symp. and Exhib.*, *34, Tomorrow's Materials: Today*, edited by G.A. Zakrzewski, D. Mazenko, S.T. Peters, and C.D. Dean, 1989, pp. 2559–2568.
15. W.W. Wright, *Compos. Polym.* **3**(4), 231–257 (1990).
16. P.W. Yip and S.S. Lin, in *Mater. Res. Soc. Symp. Proc.*, *Vol. 170, Interfaces Compos.*, edited by C.G. Pantano and E.J.H. Chen, 1990, pp. 339–344.
17. T.C. Chang and B.Z. Jang, in *Mater. Res. Soc. Symp. Proc.*, *Vol. 170 Interfaces Compos.*, edited by C.G. Pantano and E.J.H. Chen, 1990, pp. 321–326.
18. T.R. King, D.F. Adams, and D.A. Buttry, *Composites* **22**(5), 380–387 (1991).
19. M.J. Rich and L.T. Drzal, *J. Reinf. Plast. Compos.* **7**(2), 145–154 (1988).
20. B.-W. Chun, C.R. Davis, Q. He and R.R. Gustafson, *Carbon* **30**(2), 177–187 (1992).

21. L.T. Drzal, *Vacuum* **41**(7–9), 1615–1618 (1990).
22. *Adhesion and Bonding in Composites*, edited by R. Yosomiya, K. Morimoto, A. Nakajima, Y. Ikada, and T. Suzuki, Marcel Dekker, New York, 1990, pp. 257–281. (Chapter on Interfacial Effect of Carbon-Fiber-Reinforced Composite Material.)
23. *Adhesion and Bonding in Composites*, edited by R. Yosomiya, K. Morimoto, A. Nakajima, Y. Ikada, and T. Suzuki, Marcel Dekker, New York, 1990, pp. 109–154. (Chapter on Modification of Inorganic Fillers for Composite Materials.)
24. G. Sugerman, S.M. Gabayson, W.E. Chitwood, and S.J. Monte, in *Proc. 3rd Dev. Sci. Technol. Compos. Mater., Eur. Conf. Compos. Mater.*, edited by A.R. Bunsell, P. Lamicq, and A. Massiah, Elsevier, London, U.K., 1989, pp. 51–56.
25. K.T. Kern, E.R. Long, Jr., S.A.T. Long, and W.L. Harries, *Polym. Prepr. (Am. Chem. Soc., Div. Polym. Chem.)* **31**(1), 611–612 (1990).
26. W.W. Wright, *Compos. Polym.* **3**(5), 360–401 (1990).
27. R.V. Subramanian, A.R. Sanadi, and A. Crasto, *J. Adhes. Sci. Technol.* **4**(10), 329–346 (1990).
28. S.H. Jao and F.J. McGarry, in *Proc. Int. SAMPE Tech. Conf.*, 22, *Advanced Materials: Looking Ahead to the 21st Century*, edited by L.D. Michelove, R.P. Caruso, P. Adams, and W.H. Fossey, Jr., 1990, pp. 455–469.
29. C. Sellitti, J.L. Koenig, and H. Ishida, *Mater. Sci. Eng.* **A126**, 235–244 (1990).
30. B. Barbier, J. Pinson, G. Desarmot, and M. Sanchez, *J. Electrochem. Soc.* **137**(6), 1757–1764 (1990).
31. L.T. Drzal, *Mater. Sci. Eng.* **A126**, 289–293 (1990).
32. L.T. Drzal, *Mater. Res. Soc. Symp. Proc. Vol. 170, Interfaces Compos.*, edited by C.G. Pantano and E.J.H. Chen, 1990, pp. 275–283.
33. L.T. Drzal, *Treatise on Adhesion and Adhesives*, Vol. 6, edited by R.L. Patrick, Marcel Dekker: New York, 1989, pp. 187–211.
34. N.C.W. Judd, *Br. Polym. J.* **9**(4), 272–277 (1977).
35. P.J. Ives, and D.J. Williams, in *Proc. Int. SAMPE Symp. and Exhib.*, 33, *Materials: Pathway to the Future*, edited by G. Carrillo, E.D. Newell, W.D. Brown, and P. Phelan, 1988, pp. 858–869.
36. G.M. Knoblach, in *Proc. Int. SAMPE Symp. and Exhib. 34, Tomorrow's Materials: Today*, edited by G.A. Zakrzewski, D. Mazenko, S.T. Peters, and C.D. Dean, 1989, pp. 385–396.
37. E. Fitzer, in *Carbon Fibers Filaments and Composites*, edited by J.L. Figueiredo, C.A. Bernardo, R.T.K. Baker, and K.H. Huttinger, Kluwer Academic, Dordrecht, 1990, pp. 169–219.
38. W.R. Stabler, G.B. Tatterson, R.L. Sadler, and A.H.M. El-Shiekh, *SAMPE Q.* **23**(2), 38 (1992).
39. K.C. Ik and K.U. Gon, in *Proc. 7th Int. Conf. on Composite Materials, Guangzhou, China, Nov. 1989, Vol. 2*, edited by Y. Wu, Z.-L. Gu, and R. Wu, International Academic Publishers, Beijing, P.R. China, and Pergamon, Oxford, U.K. and New York, 1989, pp. 101–109.
40. O. Dickman, K. Lindersson, and L. Svensson, *Plastics and Rubber Processing and Applications* **13**, 9–14 (1990).
41. S.D. Copeland, J.C. Seferis, and M. Carrega, *J. Appl. Polym. Sci.* **44**(1), 41–53 (1992).
42. T. Hartness, in *Proc. Int. SAMPE Symp. and Exhib.*, 33, *Materials: Pathway to the Future*, edited by G. Carrillo, E.D. Newell, W.D. Brown, and P. Phelan, 1988, pp. 1458–1471.

43. H. Kosuda, Y. Nagata, and Y. Endoh, U.S. Patent 4,897,286 (1990).
44. F.K. Ko, J.-N. Chu and C.T. Hua, *J. Appl. Polym. Sci., Appl. Polym. Symp.* **47**, 501–519 (1991).
45. F. Ko, P. Fang, and H. Chu, in *Proc. Int. SAMPE Symp. and Exhib., 33, Materials: Pathway to the Future*, edited by G. Carrillo, E.D. Newell, W.D. Brown, and P. Phelan, 1988, pp. 899–911.
46. R. Weiss, *Cryogenics* **31**(4), 319–322 (1991).
47. T. Vu-Khanh and J. Denault, in *Proc. American Society for Composites, 6th Tech. Conf.*, Technomic, Lancaster, 1991, pp. 473–482.
48. J.D. Muzzy, X. Wu and J.S. Colton, *Polym. Compos.* **11**(5), 280–285 (1990).
49. A.K. Miller, M. Gur and A. Peled, *Materials & Manufacturing Processes* **5**(2), 273–300 (1990).
50. A.K. Miller, C. Chang, A. Payne, M. Gur, E. Menzel, and A. Peled, *SAMPE J.* **26**(4), 37–54 (1990).
51. A. Farouk and T.H. Kwon, *Polym. Compos.* **11**(6), 379–386 (1990).
52. J. Wei, M.C. Hawley, J. Jow, and J.D. DeLong, *SAMPE J.* **27**(1), 33–39 (1991).
53. K.J. Hook, R.K. Agrawal, and L.T. Drzal, *J. Adhes.* **32**(2–3), 157–170 (1990).
54. S.P. Grossman and M.F. Amateau, in *Proc. Int. SAMPE Symp. and Exhib., 33, Materials: Pathway to the Future*, edited by G. Carrillo, E.D. Newell, W.D. Brown, and P. Phelan, 1988, pp. 681–692.
55. J.M. Barton, J.R. Lloyd, A.A. Goodwin, and J.N. Hay, *Br. Polymer J.* **23**, 101–109 (1990).
56. S. Saiello, J. Kenny, and L. Nicolais, *J. Mater. Sci.* **25**, 3493–3496 (1990).
57. C.-C.M. Ma, S.-W. Yur, C.-L. Ong, and M.-F. Sheu, in *Proc. Int. SAMPE Symp. and Exhib., 34, Tomorrow's Materials: Today*, edited by G.A. Zakrzewski, D. Mazenko, S.T. Peters, and C.D. Dean, 1989, pp. 350–361.
58. *Sprechsaal* **123**(4), 403–408 (1990).
59. J. Iroh, J.P. Bell, and D.A. Scola, *J. Appl. Polym. Sci.* **41**(3–4), 735–749 (1990).
60. P.B. Jana and S.K. De, *Plastics, Rubber and Composites Processing and Applications* **17**. 43–49 (1992).
61. X. Shui and D.D.L. Chung, in *Proc. Int. SAMPE Symp. and Exhib., 38, Advanced Materials: Performance through Technology Insertion*, edited by V. Bailey, G.C. Janicki, and T. Haulik, 1993, pp. 1869–1875.
62. *Hercules Composite Structures*, Hercules Inc.
63. Hercules, Inc., product data, 847–4.
64. C. Blair and J. Zakrzewski, in *Proc. 22nd Int. SAMPE Tech. Conf.*, 1990, pp. 918–931.
65. E.M. Silverman, and R.J. Jones, in *Proc. Int. SAMPE Symp. and Exhib., 33, Materials: Pathway to the Future*, edited by G. Carrillo, E.D. Newell, W.D. Brown and P. Phelan, 1988, pp. 1418–1432.
66. W. Huettner and R. Weiss, in *Carbon Fibers Filaments and Composites*, edited by J.L. Figueiredo, C.A. Bernardo, R.T.K. Baker, and K.J. Huttinger, Kluwer Academic, Dordrecht, 1990, pp. 221–243.
67. J.J. Scabbo, Jr. and N. Nakajima, *SAMPE J.* **26**(1), 45–50 (1990).
68. N. Odagiri, H. Kishi, and T. Nakae, in *Proc. American Society for Composites, 6th Tech. Conf.*, Technomic, Lancaster, 1991.
69. H.H. Shim, O.K. Kwon, and J.R. Youn, *Polym. Compos.* **11**(6), 337–341 (1990).
70. P.B. Mody, T.-W. Chou, and K. Friedrich, *ASTM STP 1003, Test Methods and Design Allowables for Fibrous Composites: 2nd Volume*, edited by C.C. Chamis, 1989, pp. 75–89.

71. T.C. Ovaert and H.S. Cheng, *Wear* **150** (1–2), 275–287 (1991).
72. D.R. Rourk, in *Proc. Int. SAMPE Electron. Conf., 4, Electron. Mater.: Our Future*, 1990, pp. 167–178.
73. B. Nysten and J.-P. Issi, *Composites (Guildford, U.K.)* **21**(4), 339–343 (1990).
74. W. de la Torre, in *Proc. 6th Int. SAMPE Electron. Conf.*, 1992, pp. 720–733.
75. M. Takeno, S. Nishijima, T. Okada, K. Fujioka, Y. Tsuchida and Y. Kuraoka, *Adv. Cryog. Eng.* **32**, 217–224 (1986).
76. R.E. Schramm and M.B. Kasen, *Mater. Sci. Eng.* **30**, 197–204 (1977).
77. L. Li and D.D.L. Chung, in *Proc. Int. SAMPE Electron. Conf., 4, Electron. Mater: Our Future*, 1990, pp. 777–785.
78. P.K. Pramanik, D. Khastgir, S.K. De, and T.N. Saha, *J. Mater. Sci.* **25**, 3848–3853 (1990).
79. P.K. Pramanik, D. Khastgir, and T.N. Saha, *Composites* **23**(3), 183–191 (1992).
80. H.C. Kim and S.K. See, *J. Phys. D: Appl. Phys.* **23**(7), 916–921 (1990).
81. M.K. Abdelazeez and M.S. Ahmad, *Mater. Sci. Eng.* **B7**, 43–48 (1990).
82. G. Yaniv, G, Peimanidis, and I.M. Daniel, *ASTM STP 1003, Test Methods and Design Allowables for Fibrous Composites: 2nd Volume*, edited by C.C. Chamis, 1989, pp. 16–30.
83. D.W. Sohn and N. Sung, *Polym. Mater. Sci. Eng.* **63**, 43–47 (1990).
84. S.V. Ramani, in *Proc. 2nd Int. Conf. on Composite Materials*, 1978, p. 1602.
85. C.-C.M. Ma and S.-W. Yur, *Polym. Eng. Sci.* **31**(1), 34–39 (1991).
86. R.E. Mauri, F.W. Crossman, and W.J. Warren, in *Proc. Natl. SAMPE Symp. and Exhib.* **23**, 1978, pp. 1202–1217.
87. B.M. Parker, *Int. J. Adhes. Adhes.* **10**(3), 187–191 (1990).
88. D.A. Scola and J.H. Vontell, *Polym. Eng. Sci.* **31**(1), 6–13 (1991).
89. K.J. Bowles and G. Nowak, *J. Compos. Mater.* **22**(10), 966–985 (1988).
90. R.H. Pater, *Polym. Eng. Sci.* **31**(1), 14–19 (1991).
91. L. Li, *Polym. Plast. Technol. Eng.* **29**(5–6), 547–562 (1990).
92. A. Guemes and W. Barrera, in *Proc. 7th Int. Conf. on Composite Materials, Guangzhou, China, Nov. 1989, Vol. 2*, edited by Y. Wu, Z. Gu, and R. Wu, International Academic Publishers, Beijing, P.R. China, and Pergamon, Oxford, U.K. and New York, 1989, pp. 222–225.
93. I. Gilath, S. Eliezer, and S. Shkolnik, *J. Compos. Mater.* **24**(11), 1138–1151 (1990).
94. W.C. Tucker, *PD (Am. Soc. Mech. Eng.)*, **32** (Compos. Mater. Technol.), 95–99 (1990).
95. M. Dole, *Radiat. Phys. Chem.* **37**(1), 65–70 (1991).
96. W.J. Cantwell, P. Davies, P.-E. Bourban, P.-Y. Jar, and H.H. Kausch, *Swiss. Mater.* **2**(1), 25–29 (1990).
97. E.M. Silverman, R.A. Griese, and W.F. Wright, in *Proc. Int. SAMPE Symp. and Exhib., 34, Tomorrow's Materials: Today*, edited by G.A. Zakrzewski, D. Mazenko, S.T. Peters, and C.D. Dean, 1989, pp. 770–779.
98. W.J. Cantwell, P. Davies, and H.H. Kausch, *SAMPE J.* **27**(6), 30–35 (1991).
99. J.C. Domanus and H. Lilholt, in *Proc. 2nd Int. Conf. on Composite Materials*, 1978, TMS–AIME, pp. 1072–1092.
100. M.J. Folkes and H.A. Potts, *Plast. Rubber Process. Appl.* **10**(2), 79–84 (1988).
101. D.R. Dunbar, A.R. Robertson, and R. Kerrison, in *Proc. 2nd Int. Conf. on Composite Materials*, 1978, TMS–AIME, pp. 1360–1375.

102. M.G. Hammond and K. Farrell, in *Proc. 2nd Int. Conf. on Composite Materials*, 1978, TMS–AIME, pp. 1392–1404.
103. A. Knoell and G. Krumweide, in *Proc. 2nd Int. Conf. on Composite Materials*, 1978, TMS–AIME, pp. 1377–1391.
104. J.W. Young and T.A. Dougherty, *Acta Astronaut.* **4**(7–8), 833–846 (1977).
105. R.F. Siegmund, *Chimia* **44**(11), 361–364 (1990).
106. E.M. Silverman and R.J. Jones, in *Proc. Int. SAMPE Symp. and Exhib., 33, Materials: Pathway to the Future*, edited by G. Carrillo, E.D. Newell, W.D. Brown, and P. Phelan, 1988, pp. 1418–1432.
107. A. Thorne and L. Hollaway, European Space Agency Special Publication ESA SP, ESA SP-303, pp. 207–211 (1990).
108. A. Barrio Cardaba, F. Rodríguez Lence, and J. Sánchez Gómez, *SAMPE J.* **26**(1), 9–13 (1990).
109. S.M. Lee, T. Jonas, and G. DiSalvo, *SAMPE J.* **27**(2), 19–25 (1991).
110. N. Waterman, *Engineer*, September 1978, pp. 48–54.
111. D.N. Yates, J. Presta, D. Sidwell, P. Sharifi, and R. Torczyner, in *Proc. 2nd Int. Conf. on Composite Materials*, 1978, TMS–AIME, p. 1445.
112. T. Akasaka, M. Masutani, T. Nakakura and H. Sakai, in *Proc. Int. SAMPE Symp. and Exhib., 33, Materials: Pathway to the Future*, edited by G. Carrillo, E.D. Newell, W.D. Brown, and P. Phelan, 1988, pp. 670–680.
113. B. Ballance, *Carbon Fibers*, edited by The Plastics and Rubber Inst., London, England, Noyes, Park Ridge, NJ, 1986, pp. 72–85.
114. H. Hillesland, J. Holbery and T. SeCoy, in *Proc. Int. SAMPE Symp. and Exhib., 34, Tomorrow's Materials: Today*, edited by G.A. Zakrzewski, D. Mazenko, S.T. Peters, and C.D. Dean, 1989, pp. 1571–1577.
115. J.T. Kung and M.P. Amason, in *Proc. Natl. SAMPE Symp. and Exhib., 23*, 1978, pp. 1039–1053.
116. M.A. Council and G.B. Park, in *Proc. Int. SAMPE Symp. and Exhib., 34, Tomorrow's Materials: Today*, edited by G.A. Zakrzewski, D. Mazenko, S.T. Peters and C.D. Dean, 1989, pp. 1644–1655.
117. G.R. Stafford, G.L. Cahen, Jr., and G.E. Stoner, *J. Electrochem. Soc.* **138**(2), 425–430 (1991).
118. L. Nacamulli and E. Gileadi, *J. Appl. Electrochem.* **12**, 73–78 (1982).
119. A.M. Ibrahim, in *Proc. 6th Int. SAMPE Electron. Conf.*, 1992, pp. 556–567.
120. D. Maass and M. Makwinski, in *Proc. 6th Int. SAMPE Electron. Conf.*, 1992, pp. 578–593.
121. B.R. Lyons and M. Molyneux, in *Proc. 2nd Int. Conf. on Composite Materials*, 1978, TMS–AIME, pp. 1474–1492.
122. M.C. Zimmerman, H. Alexander, J.R. Parsons, and P.K. Bajpai, *ACS Symp. Ser., Vol. 457, High-Tech. Fibrous Materials*, American Chemical Society, Washington, D.C., 1991, pp. 132–148.
123. S.A. Brown, R.S. Hastings, J.J. Mason, and A. Moet, *Biomaterials* **11**(8), 541–547 (1990).
124. M. Spector, E.J. Cheal, R.D. Jamison, S. Alter, N. Madsen, L. Strait, G. Maharaj, A. Gavins, D.T. Reilly and C.B. Sledge, in *Proc. 22nd Int. SAMPE Tech. Conf.*, 1990, pp. 1119–1130.
125. T. Fujisaki, S. Kokusho, K. Kobayashi, S. Hayashi, C. Ito, and M. Arai, *Rep. Res. Lab. Eng. Mater., Tokyo Inst. Technol.*, **15**, 207–219 (1990).
126. T. Uomoto and H. Hodhod, *Seisan Kenkyu* **43**(3), 161–164 (1991).
127. Y. Suda, U.S. Patent 4,919,859 (1990).
128. S. Suzuki and K. Kato, Jpn. Kokai 77,105,959 (1977).
129. H. Kitagawa, U.K. Patent Application GB 2,224,510A (1990).

CHAPTER **7**

Metal-Matrix Composites

Introduction

Carbon fiber metal-matrix composites are gaining importance because the carbon fibers serve to reduce the coefficient of thermal expansion (Figure 7.1 [1]), increase the strength and modulus, and decrease the density. If a relatively graphitic kind of carbon fiber is used, the thermal conductivity can be enhanced also (Figure 7.2 [2]). Table 7.1 [1] shows the coefficient of thermal expansion (CTE) and thermal conductivity of carbon fiber (P-120) aluminum-matrix and copper-matrix composites compared to other materials. The thermal conductivities of both composites exceed those of their corresponding matrix materials. The CTEs of both composites are nearly zero. This combination of low CTE and high thermal conductivity makes them very attractive for electronic packaging (e.g., heat sinks). As well as good thermal

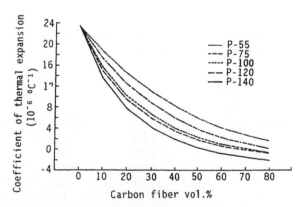

Figure 7.1   Coefficient of thermal expansion vs. carbon fiber volume percentage for aluminum reinforced with various kinds of continuous carbon fibers (Amoco's Thornel P-55, P-75, P-100, P-120, and P-140, in order of increasing fiber modulus) in a crossplied configuration. From Ref. 1. (Reprinted by permission of the Society for the Advancement of Material and Process Engineering.)

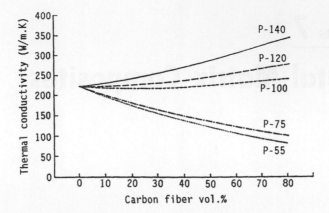

Figure 7.2 Thermal conductivity versus carbon fiber volume percentage for aluminum reinforced with various kinds of continuous carbon fibers (Amoco's Thornel P-55, P-75, P-100, P-120, and P-140, in order of increasing fiber modulus) in a crossplied configuration. From Ref. 1. (Reprinted by permission of the Society for the Advancement of Material and Process Engineering.)

properties, their low density makes them particularly desirable for aerospace electronics and orbiting space structures; orbiters are thermally cycled by moving through the earth's shadow.

Compared to the metal itself, a carbon fiber metal-matrix composite is characterized by a higher strength-to-density ratio (i.e., specific strength, Figure 7.3 [2]), a higher modulus-to-density ratio (i.e., specific modulus, Figure 7.3 [2]), better fatigue resistance, better high temperature mechanical properties (a higher strength and a lower creep rate), a lower CTE, and better wear resistance.

Table 7.1 Thermal properties of carbon fiber aluminum-matrix and copper-matrix composites compared to other materials. From Ref. 1.

Material	Reinforcement (vol. %)	Density (g/cm³)	Axial thermal conductivity (W/mK)	Axial CTE ($10^{-6}/°C$)
Alumina		3.60	22.36	5.8–7.7
Copper		8.94	391	17.6
Aluminum		2.71	221	23.6
Molybdenum		10.24	146	5.2
Kovar		8.19	17	5.8
Copper/invar/copper		8.19	131	5.8
Al/SiC particles	40	2.91	128	12.6
Al/P-120 carbon fibers	60	2.41	419	−0.32
Cu/P-120 carbon fibers	60	6.23	522	−0.07

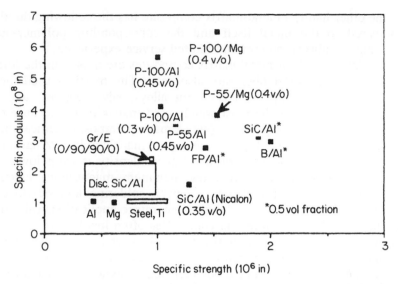

Figure 7.3 Specific stiffness (modulus) versus specific strength of aluminum- and magnesium-matrix composites, compared to the unreinforced alloys. Properties of continuous fiber reinforced materials are calculated parallel to the fibers. From Ref. 2. (Reprinted with permission from *JOM* (formerly *Journal of Metals*) Vol. 40, No. 2, 1988, a publication of The Minerals, Metals & Materials Society, Warrendale, PA 15086.)

Compared to carbon fiber polymer-matrix composites, a carbon fiber metal-matrix composite is characterized by higher temperature resistance, higher fire resistance, higher transverse strength and modulus, the lack of moisture absorption, a higher thermal conductivity, a lower electrical resistivity, better radiation resistance, and absence of outgassing. Table 7.2 shows the CTE (α), thermal conductivity (κ), Poisson's ratio (ν) and Young's modulus (E) of carbon fiber (Amoco's Thornel P-100 and P-55) reinforced aluminum (alloy 6061) and carbon fiber reinforced epoxy [3].

Table 7.2 Properties of carbon fiber reinforced aluminum (6061 alloy) compared to carbon fiber reinforced epoxy. from Ref. 3.

Property	Graphite/epoxy	P-100/6061	P-55/6061
α_{11} (10^{-5}/K)	−0.080	0.086	0.307
α_{22} (10^{-5}/K)	3.67	2.30	2.40
κ_{11} (W/m/K)	54.0	240.0	98.0
κ_{22} (W/m/K)	0.7	193.0	98.0
ν_{12}	0.21	0.4	0.27
ν_{21}	0.010	0.031	0.041
E_{11} (GPa)	172	352	213
E_{22} (GPa)	8.07	28	32
E_{12} (GPa)	4.28	14	13

On the other hand, a metal-matrix composite has the following disadvantages compared to the metal itself and the corresponding polymer-matrix composite: higher fabrication cost and limited service experience.

Carbon fibers used for metal-matrix composites are mostly in the form of continuous fibers, but short fibers are also used. The matrices used include aluminum, magnesium, copper, nickel, tin alloys, silver–copper, and lead alloys. Aluminum is by far the most widely used matrix metal because of its low density, low melting temperature (which makes composite fabrication and joining relatively convenient), low cost, and good machinability. Magnesium is comparably low in melting temperature, but its density is even lower than aluminum. Applications include structures (aluminum, magnesium), electronic heat sinks and substrates (aluminum, copper), soldering and bearings (tin alloys), brazing (silver–copper), and high-temperature applications (nickel).

The fabrication of metal-matrix composites often involves the use of an intermediate, called a preform, in the form of sheets, wires, cylinders, or near-net shapes. The preform contains the reinforcing fibers usually held together by a binder, which can be a polymer (e.g., acrylic, styrene), a ceramic (e.g., silica, aluminum metaphosphate [4]), or the matrix metal itself. For example, continuous fibers are wound around a drum and bound with a resin, and subsequently the wound fiber cylinder is cut off the drum and stretched out to form a sheet. During subsequent composite fabrication, the organic binder evaporates. As another example, short fibers are combined with a ceramic or polymeric binder and a liquid carrier (usually water) to form a slurry; this is then filtered under pressure or wet pressed to form a wet "cake," which is subsequently dried to form a preform. In the case of using the matrix metal as the binder, a continuous fiber bundle is immersed in the molten matrix metal so as to be infiltrated with it, thus forming a wire preform; alternately, fibers placed on a matrix metal foil are covered and fixed in place with a sprayed matrix metal, thus forming a sprayed preform.

A binder is not always needed, although it helps the fibers to stay uniformly distributed during subsequent composite fabrication. Excessive ceramic binder amounts should be avoided, as they can make the resulting metal-matrix composite more brittle. For ceramic binders, a typical amount ranges from 1 to 5 wt.% of the preform [4]. In the case of woven fabrics as reinforcement, a binder is less important, as the weaving itself serves to hold the fibers together in a uniform fashion.

Fabrication

The most popular method for the fabrication of carbon fiber metal-matrix composites is the infiltration of a preform by a liquid metal under pressure. The low viscosity of liquid metals compared to resins or glasses makes infiltration very appropriate for metal-matrix composites. Nevertheless, pressure is required because of the difficulty for the liquid metal to wet the carbon fibers. The pressure can be provided by a gas (e.g., argon), as illustrated in

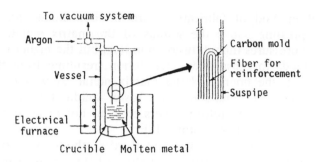

Figure 7.4 Schematic view of an apparatus for liquid metal infiltration under vacuum and inert gas pressure. From Ref. 6. (By permission of the publishers, Butterworth–Heinemann Ltd.)

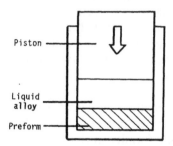

Figure 7.5 Schematic view of squeeze casting, i.e., direct infiltration of a preform, using a piston to provide pressure.

Figure 7.4, or a piston, as illustrated in Figure 7.5. When a piston is used, the process can be quite fast and is known as squeeze casting.

A second method for fabricating carbon fiber metal-matrix composites is diffusion bonding. In this method, a stack of alternating layers of carbon fibers and metal foils is hot pressed (at, say, 24 MPa for 20 min.) to cause bonding in the solid state. This method is not very suitable for fiber cloths or continuous fiber bundles because of the difficulty for the metal to flow to the space between the fibers during diffusion bonding. In contrast, the infiltration method involves melting the metal, so metal flow is relatively easy, making infiltration a more suitable method for fiber cloths or continuous fiber bundles. A variation of diffusion bonding involves the hot pressing of metal-coated carbon fibers without the use of metal foils. In this case, the metal coating provides the metal for the metal-matrix composite. In general, the diffusion bonding method is complicated by the fact that the surface of the metal foil or metal coating tends to be oxidized and the oxide makes the bonding more difficult. Hence, a vacuum is usually required for diffusion bonding.

A third method of fabricating carbon fiber metal-matrix composites involves hot pressing above the solidus of the matrix metal. This method requires lower pressures than diffusion bonding, but the higher temperature of the pressing tends to cause fiber degradation, resulting from the interfacial reaction between the fibers and the matrix metal. A way to alleviate this problem is to insert a metal sheet of a lower solidus than the matrix metal alternately between wire preform layers then to hot press at a temperature between the two solidus temperatures [5].

A combination of the second and third methods involves first heating carbon fibers laid up with matrix metal sheets between them in vacuum in a sealed metal container above the liquidus of the matrix metal, then immediately hot pressing the container at a temperature below the solidus of the matrix metal [5].

A fourth method of fabricating carbon fiber metal-matrix composites involves the plasma spraying of the metal on to continuous fibers. As this process usually results in a composite of high porosity, subsequent consolidation (e.g., by hot isostatic pressing) is usually necessary. Compared to the other methods, plasma spraying has the advantage of being able to produce continuous composite parts, though the subsequent consolidation step may limit their size.

Slurry casting, a fifth method, is complicated by the tendency for the carbon fibers (low in density) to float on the metal melt. To overcome this problem, which causes nonuniformity in the fiber distribution, compocasting is necessary. Compocasting (rheocasting) involves vigorously agitating a semisolid alloy so that the primary phase is nondendritic, thereby giving a fiber–alloy slurry with thixotropic properties.

The infiltration method can be used to produce near-net shape composites, so that subsequent shaping is not necessary. As infiltration using a gas pressure involves a smaller rate of pressure increase compared to using a piston, infiltration is more suitable than squeeze casting for near-net shape processing. If shaping is necessary, it can be achieved by plastic forming (e.g., extrusion, swaging, forging, and rolling) for the case of short fibers. Plastic forming tends to reduce the porosity and give a preferential orientation to the short fibers. These effects result in improved mechanical properties. For continuous fiber metal-matrix composites, shaping cannot be achieved by plastic forming and cutting is necessary.

Wetting of Carbon Fibers by Molten Metals

The difficulty for molten metals to wet the surface of carbon fibers complicates the fabrication of the metal-matrix composites. This difficulty is particularly severe for high-modulus carbon fibers (e.g., Amoco's Thornel P-100) which have graphite planes mostly aligned parallel to the fiber surface. The edges of the graphite planes are more reactive with the molten metals than the graphite planes themselves, low-modulus carbon fibers are more reactive and thus are wetted more easily by the molten metals. Although this reaction between the fibers and the metal helps the wetting, it produces a brittle carbide and degrades the strength of the fibers.

In order to enhance the wetting, carbon fibers are coated by a metal or a ceramic. The metals used as coatings are formed by plating and include Ni, Cu, and Ag; they generally result in composites of strengths much lower than those predicted by the rule of mixtures (ROM). In the case of nickel-coated and copper-coated carbon fibers in an aluminum matrix, metal aluminides (Al_3M) form and embrittle the composites. In the case of nickel-coated carbon fibers in a magnesium matrix, nickel reacts with magnesium to form Ni/Mg compounds and a low melting (508°C) eutectic [6]. On the other hand, copper-coated fibers are suitable for copper [7], tin, or other metals as the matrix. A metal coating that is particularly successful involves sodium, which wets carbon fibers and coats them with a protective intermetallic compound by reaction with one or more other molten metals (e.g., tin). This is called the sodium process [8,9]. A related process immerses the fibers in liquid NaK [10]. However, these processes involving sodium suffer from sodium contamination of the fibers, probably due to the intercalation of sodium into graphite [11]. Nevertheless, aluminum-matrix composites containing unidirectional carbon fibers treated by the sodium process exhibit tensile strengths close to those calculated by using the rule of mixtures, indicating that the fibers are not degraded by the sodium process [11].

Examples of ceramics used as coatings on carbon fibers are TiC, SiC, B_4C, TiB_2, TiN, K_2ZrF_6, ZrO_2. Methods used to deposit the ceramics include (1) reaction of the carbon fibers with a molten metal alloy, called the liquid metal transfer agent (LMTA) technique, (2) chemical vapor deposition (CVD), and (3) solution coating.

The LMTA technique involves immersing the fibers in a melt of copper or tin (called a liquid metal transfer agent, which must not react with carbon) in which a refractory element (e.g., W, Cr. Ti) is dissolved, and subsequent removal of the transfer agent from the fiber surface by immersion in liquid aluminum (suitable for fabricating an aluminum-matrix composite). For example, for forming a TiC coating, the alloy can be Cu–10% Ti at 1 050°C or Sn–1% Ti at 900–1 055°C. In particular, by immersing the fibers in Sn–1% Ti at 900–1 055°C for 0.25–10 min., a 0.1 μm layer of TiC is formed on the fibers, although they are also surrounded by the tin alloy. Subsequent immersion for 1 min. in liquid aluminum causes the tin alloy to dissolve in the liquid aluminum [12]. The consequence is a wire preform suitable for fabricating aluminum-matrix composites. Other than titanium carbide, tungsten carbide and chromium carbide have been formed on carbon fibers by the LMTA technique.

The CVD technique has been used for forming coatings of TiB_2, TiC, SiC, B_4C and TiN. The B_4C coating is formed by reactive CVD on carbon fibers, using a BCl_3/H_2 mixture as the reactant [13]. The TiB_2 deposition uses $TiCl_4$ and BCl_3 gases, which are reduced by Zn vapor, as illustrated in Figure

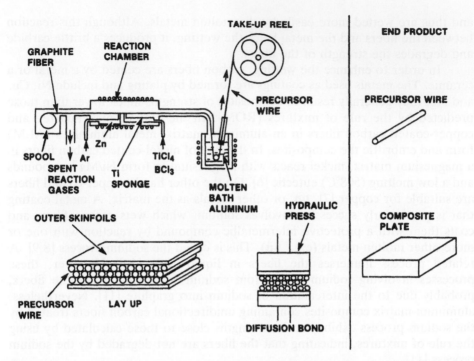

Figure 7.6 Schematic of aluminum-matrix composite fabrication by diffusion bonding of precursor wires in the form of aluminum-impregnated TiB_2-coated carbon fibers. From Ref. 14. (Reprinted by courtesy of Marcel Dekker, Inc.)

7.6. The TiB_2 coating is particularly attractive because of the exceptionally good wetting between TiB_2 and molten aluminum. During composite fabrication, the TiB_2 coating is displaced and dissolved in the matrix, while an oxide (γ-Al_2O_3 for a pure aluminum matrix, $MgAl_2O_4$ spinel for a 6061 aluminum matrix) is formed between the fiber and the matrix. The oxygen for the oxide formation comes from the sizing on the fibers; the sizing is not completely removed from the fibers before processing [14]. The oxide layer serves as a diffusion barrier to aluminum, but allows diffusion of carbon, thereby limiting Al_4C_3 growth to the oxide–matrix interface [15]. Moreover, the oxide provides bonding between the fiber and the matrix. Because of the reaction at the interface between the coating and the fiber, the fiber strength is degraded after coating. To alleviate this problem, a layer of pyrolytic carbon is deposited between the fiber and the ceramic layer [16]. The CVD process involves high temperatures, e.g., 1 200°C for SiC deposition using CH_3SiCl_4 [17]; this high temperature degrades the carbon fibers. Another problem of the CVD process is the difficulty of obtaining a uniform coating around the circumference of each fiber. Moreover, it is expensive and causes the need to scrub and dispose most of the corrosive starting material, as most of the starting material does not react at all. The most serious problem with the TiB_2 coating is that it is not

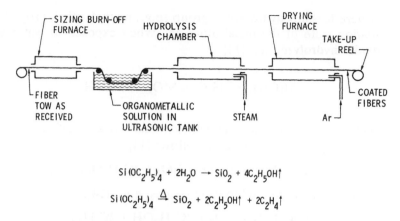

$$Si(OC_2H_5)_4 + 2H_2O \rightarrow SiO_2 + 4C_2H_5OH\uparrow$$

$$Si(OC_2H_5)_4 \xrightarrow{\Delta} SiO_2 + 2C_2H_5OH\uparrow + 2C_2H_4\uparrow$$

Figure 7.7 Organometallic solution coating process for carbon fibers. From Ref. 14. (Reprinted by courtesy of Marcel Dekker, Inc.)

air stable; it cannot be exposed to air before immersion in the molten metal otherwise wetting will not take place. This problem limits the shape of materials that can be fabricated, especially since the wire preforms are not very flexible [14].

A high compliance (or a low modulus) is preferred for the coating in order to increase the interface strength. An increase in the interface (or interphase) strength results in an increase in the transverse strength. The modulus of SiC coatings can be varied by controlling the plasma voltage in plasma assisted chemical vapor deposition (PACVD). Modulus values in the range from 19 to 285 GPa have been obtained in PACVD SiC, compared to a value of 448 GPa for CVD SiC. Unidirectional carbon fiber (Thornel P-55) aluminum-matrix composites in which the fibers are coated with SiC exhibit an interface strength and a transverse strength which increase with decreasing modulus of the SiC coating [18–20].

The most attractive coating technique developed to date is the solution coating method. In the case of using an organometallic solution, fibers are passed through a toluene solution containing an organometallic compound, followed by hydrolysis or pyrolysis of the organometallic compounds to form the coating. Thus, the fibers are passed sequentially through a furnace in which the sizing on the fibers is vaporized, followed by an ultrasonic bath containing an organometallic solution. The coated fibers are then passed through a chamber containing flowing steam in which the organometallic compound on the fiber surface is hydrolyzed to oxide and finally through an argon atmosphere drying furnace in which any excess solvent or water is vaporized and any unhydrolyzed organometallic is pyrolyzed [14]. The process is illustrated in Figure 7.7. In contrast to the TiB_2 coatings, the SiO_2 coatings formed by organometallic solution coating are air stable.

The organometallic compounds used are alkoxides, in which metal atoms are bound to hydrocarbon groups by oxygen atoms. The general formula is

M(OR)$_x$, where R is any hydrocarbon group (e.g., methyl, ethyl, propyl) and x is the oxidation state of the metal atom M. When exposed to water vapor, these alkoxides hydrolyze, i.e., [14]:

$$M(OR)_x + \frac{x}{2}H_2O \rightarrow MO_{x/2} + xROH$$

For example, the alkoxide tetraethoxysilane (also called tetraethylorthosilicate) is hydrolyzed by water as follows [14]:

$$Si(OC_2H_5)_4 + 2H_2O \rightarrow SiO_2 + 4C_2H_5OH$$

Alkoxides can also be pyrolyzed to yield oxides, e.g., [14]:

$$Si(OC_2H_5)_4 \rightarrow SiO_2 + 2C_2H_5OH + 2C_2H_4$$

Most alkoxides can be dissolved in toluene. By controlling the solution concentration and the time and temperature of immersion, it is possible to control the uniformity and thickness of the resulting oxide coating. The thickness of the oxide coatings on the fibers varies from 700 to 1 500 Å. The oxide is amorphous and contains carbon, which originates in the carbon fiber. The elemental concentration profiles obtained by Auger depth profiling are shown in Figure 7.8 [14].

Liquid magnesium wets SiO$_2$-coated low-modulus carbon fibers (e.g., T-300) and infiltrates the fiber bundles, due to reactions between the molten magnesium and the SiO$_2$ coating. The reactions include the following [14].

$$2Mg + SiO_2 \rightarrow 2MgO + Si \qquad \Delta G°_{670°C} = -76\,kcal$$

$$MgO + SiO_2 \rightarrow MgSiO_3 \qquad \Delta G°_{670°C} = -23\,kcal$$

$$2Mg + 3SiO_2 \rightarrow 2MgSiO_3 + Si \qquad \Delta G°_{670°C} = -122\,kcal$$

$$2MgO + SiO_2 \rightarrow Mg_2SiO_4 \qquad \Delta G°_{670°C} = -28\,kcal$$

$$2Mg + 2SiO_2 \rightarrow Mg_2SiO_4 + Si \qquad \Delta G°_{670°C} = -104\,kcal$$

The interfacial layer between the fiber and the Mg matrix contains MgO and magnesium silicates. However, immersion of SiO$_2$-coated high-modulus fibers (e.g., P-100) in liquid magnesium causes the oxide coating to separate from the fibers, due to the poor adherence of the oxide coating to the high modulus fibers. This problem with the high-modulus fibers can be solved by first depositing a thin amorphous carbon coating on the fibers by passing the fiber bundles through a toluene solution of petroleum pitch, followed by evaporation of the solvent and pyrolysis of the pitch, as illustrated in Figure 7.9 [14].

The most effective air-stable coating for carbon fibers used in aluminum-matrix composites is a mixed boron–silicon oxide applied from organometallic solutions [14].

Instead of SiO$_2$, TiO$_2$ can be deposited on carbon fibers by the

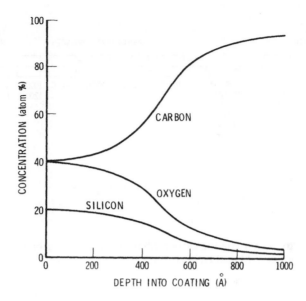

Figure 7.8 Concentration profiles of the elements in SiO_2 coatings on carbon fibers. From Ref. 14. (Reprinted by courtesy of Marcel Dekker, Inc.)

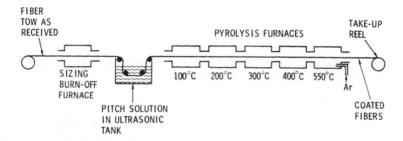

Figure 7.9 Process for coating carbon fibers with amorphous carbon. From Ref. 14. (Reprinted by courtesy of Marcel Dekker, Inc.)

organometallic solution method. For TiO_2, the alkoxide can be titanium isopropoxide [21].

SiC coatings can be formed by using polycarbosilane (dissolved in toluene) as the precursor, which is pyrolyzed to SiC. The process for coating carbon fibers with SiC is illustrated in Figure 7.10. These coatings are wet by molten copper containing a small amount of titanium, due to a reaction between SiC and Ti to form TiC [14].

Instead of using an organometallic solution, another kind of solution coating method uses an aqueous solution of a salt. For example, the salt potassium zirconium hexafluoride (K_2ZrF_6) or potassium titanium hexafluoride

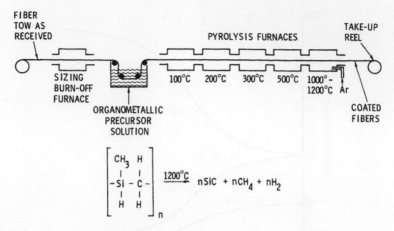

Figure 7.10 Process for coating carbon fibers with silicon carbide. From Ref. 14. (Reprinted by courtesy of Marcel Dekker, Inc.)

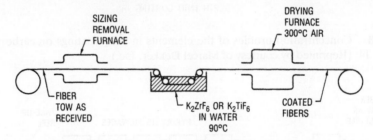

Figure 7.11 Process for coating carbon fibers with a fluoride (K_2ZrF_6 or K_2TiF_6). From Ref. 14. (Reprinted by courtesy of Marcel Dekker, Inc.)

(K_2TiF_6) is used to deposit microcrystals of K_2ZrF_6 or K_2TiF_6 on the fiber surface [14,22,23]. These fluoride coatings are stable in air. A schematic of the coating process is shown in Figure 7.11 [14]. The following reactions supposedly take place [22] between K_2ZrF_6 and the aluminum matrix:

$$3K_2ZrF_6 + 4Al \rightarrow 6KF + 4AlF_3 + 3Zr \tag{1}$$

$$3Zr + 9Al \rightarrow 3Al_3Zr \tag{2}$$

$$Zr + O_2 \rightarrow ZrO_2 \tag{3}$$

In the case of an Al–12 wt.% Si alloy (rather than pure Al) as the matrix, the following reaction may also occur [23]:

$$Zr + 2Si \rightarrow ZrSi_2 \tag{4}$$

The fluorides KF and AlF_3 are thought to dissolve the thin layer of Al_2O_3 which is on the liquid aluminum surface, thus helping the liquid aluminum to wet the carbon fibers. Furthermore, the reactions (1) and (2) are strongly

Table 7.3 Wettability of aluminum binary alloys on carbon fibers. From Ref. 26.

Alloying element	Addition (at.%)	Wettability
Mn	0.5	Poor
Mg	1.0	Poor
	5.0	Better
Ge	1.0	Poor
	2.3	Poor
Cr	0.5	Poor
Cu	2.5	Poor
Si	1.0	Poor
	5.0	Better
Ga	1.0	Poor
Sn	1.0	Poor
	5.0	Poor
Pb	1.0	Good
In	1.0	Good
Tl	1.0	Good

exothermic and may therefore cause a local temperature increase near the fiber–matrix interface. The increased temperature probably gives rise to a liquid phase at the fiber–matrix interface [22].

Although the K_2ZrF_6 treatment causes the contact angle between carbon and liquid aluminum at 700–800°C to decrease from 160° to 60–75° [22], it causes degradation of the fiber tensile strength during aluminum infiltration [24].

Another example of a salt solution coating method involves the use of the salt zirconium oxychloride ($ZrOCl_2$) [25]. Dip coating the carbon fibers in the salt solution and subsequently heating at 330°C cause the formation of a ZrO_2 coating of less than 1 μm in thickness. The ZrO_2 coating serves to improve fiber–matrix wetting and reduce the fiber–matrix reaction in aluminum-matrix composites.

Instead of treating the carbon fibers, the wetting of the carbon fibers by molten metals can be improved by the addition of alloying elements into the molten metals. For aluminum as the matrix, effective alloying elements include Mg, Cu, and Fe [11]. Table 7.3 [26] lists the wettability of various binary aluminum alloys on carbon fibers.

Even though the surfaces of a fiber bundle are wet well with the molten metal, infiltration of the metal into the fiber bundle is limited. This problem can be alleviated by ultrasonic vibration of the molten metal, as illustrated in

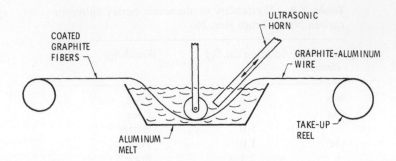

Figure 7.12 Ultrasonic vibration of an aluminum melt to assist the infiltration of a bundle of coated carbon fibers. From Ref. 14. (Reprinted by courtesy of Marcel Dekker, Inc.)

Figure 7.12 for the case of infiltrating a continuous bundle of coated carbon fibers with aluminum [14].

Degradation by Heat and Water

The degradation of carbon fiber metal-matrix composites at temperatures $\geqslant 500°C$ or in water environments is of great relevance to their practical use.

High-temperature degradation is partly due to oxidation of the carbon fibers and partly due to reaction between the carbon fibers and the matrix metal to form carbides (Al_4C_3 in the case of an aluminum-matrix composite; no reaction between magnesium and carbon in the case of a magnesium-matrix composite). Less fiber degradation due to oxidation occurs in the absence of air, but carbide formation occurs whether or not air is present. For example, heating aluminum-coated carbon fibers in a vacuum (10^{-5} torr) at 600°C for 1 h. caused the tensile strength of the carbon fibers to decrease by 13%, while similar heating at 650°C for 1 h. caused the strength to decrease by 59% [27]. Heating the composite, for the purpose of aging (precipitation hardening) the aluminum matrix, is detrimental to its mechanical properties due to the fiber–matrix reaction, as shown for the case of aging a 6061 aluminum matrix by solution treatment in air at 530–570°C, followed by aging in air at 160–320°C [28]. The extent and nature of the thermal degradation depend on the matrix alloy; for example, Al–12% Si leads to more fiber oxidation but less Al_4C_3 formation than Al–10% Mg [29].

The degradation of the composites in water environments is partly due to the hydrolysis of the metal carbide at the fiber–matrix interface and is partly due to the galvanic coupling provided by water and an electrolyte between carbon fibers and the metal matrix. For the case of an aluminum-matrix composite, the hydrolysis is of the form:

$$Al_4C_3 + 12H_2O \rightarrow 4Al(OH)_3 + 3CH_4$$

In the presence of an electrolyte solution, a corrosion cell is formed between carbon fibers and the aluminum matrix (the anode), resulting in the dissolution of the aluminum matrix to form $Al(OH)_3$. Because $Al(OH)_3$ has a higher specific volume than aluminum, the formation of $Al(OH)_3$ leads to crevices, which enhance water penetration. The hydrolysis of Al_4C_3 gives additional passages for corrosion. As the amount of Al_4C_3 increases with temperature, heating aggravates the corrosion problem [30].

The reaction between carbon fibers and the metal matrix occurs to a limited extent during composite fabrication because of the high temperatures encountered by the fibers. For the case of aluminum-matrix composites fabricated by squeeze casting (with a pour temperature of 750–850°C, a mold temperature of about 300°C, and a contact time of a few seconds between the liquid aluminum and the carbon fibers), the tensile strength of the carbon fibers is decreased by the composite fabrication by up to 12%, as shown by afterwards etching away the metal matrix [29]. Aluminum-matrix composite fabrication by solid phase diffusion bonding in a vacuum for 1 h. causes the carbon fiber strength to degrade by 13% if the bonding is performed at 600°C, and by 60% if the bonding is performed at 650°C [31].

In spite of the thermal degradation of carbon fiber aluminum-matrix composites, these composites do retain their tensile strength up to 450°C [5]. In contrast, most aluminum alloys retain their strength up to about 200°C only. For a composite containing 60 vol.% continuous unidirectional pitch-based high-modulus carbon fibers, a tensile strength of 1 500 MPa and a Young's modulus of 430 GPa have been reported [5]. In addition to increasing the tensile strength, the carbon fibers serve to increase the fatigue resistance [5] and the wear resistance [32].

In general, carbon fiber metal-matrix composites are at least as susceptible to corrosion as the unreinforced matrix [33]. For corrosion protection, carbon fiber metal-matrix composites are subjected to a surface treatment, which can be sulfuric acid anodizing, chromate/phosphate conversion coating, Al/Mn electrodeposition [34], or polymer coating [35]. Sulfuric acid anodizing forms a stable oxide coating (10–30 μm thick), followed by sealing in sodium dichromate. Chromate/phosphate conversion coatings are formed by a chemical oxidation–reduction reaction, which converts the natural aluminum oxide film to a chromate, a phosphate, or a complex oxide [34]. Al/Mn electrodeposition involves forming a very thin electroless nickel coating, followed by plating by Al/Mn [34]. Polymer coating involves the application of an epoxy/polyamide coating [35]. Sulfuric acid anodizing provides protection of aluminum-matrix composites in a marine environment for at least 180 days; chromate/phosphate conversion and electrodeposited Al/Mn coatings provide protection of aluminum-matrix composites in a marine environment for at least 365 days [34]. A 10 μm epoxy/polyamide coating provides protection of aluminum-matrix composites in a marine environment for at least 21 days; a 25 μm epoxy/polyamide coating provides protection of magnesium-matrix composites in a marine environment for at least 6 days [35]. In aerospace environments where oxygen and water are absent, the corrosion of carbon fiber metal-matrix composites is essentially not a problem.

The thermal expansion of continuous fiber metal-matrix composites is different from that of the unreinforced matrix in that the strain–temperature curve deviates from linearity above a definite transition temperature, which is highly dependent on the matrix. This nonlinearity is due to the yielding of the matrix. The yielding is a result of the internal thermal stresses [36]. The elastic limit can be increased to avoid yielding by thermal processing [37]. In general, the composites should be heat-treated prior to the measurement of the thermal expansion coefficient. Due to the thermal expansion difference between the fibers and aluminum, the thermal stresses generated by thermal cycling are considerably larger than the stress applied during thermal cycling. Therefore, the composite creeps in the reverse sense to the applied load during the cooling part of the cycle [38].

The plastic energy absorbed in the impact fracture of continuous carbon fiber aluminum-matrix composites is mainly contributed by fiber pullout. The contribution by the fiber–matrix debonding amounts to less than 10% of the fiber pullout energy [39].

Composites with Matrices other than Al and Mg

Copper has a very high thermal conductivity, together with a very low electrical resistivity. This combination of properties makes it attractive for brushes in motors and related electrical devices [40]. Moreover, the high thermal conductivity makes it attractive for electronic packaging [41–43]. However, its high density makes it less attractive for aerospace applications.

Due to the high melting temperature of copper, copper-matrix composites are not conveniently made by infiltration, but by diffusion bonding. However, higher tensile strengths that follow the rule of mixtures are obtained by infiltration [44]. Carbon fiber copper-matrix composites are most conveniently made by hot pressing copper-plated (1–3 μm thick) carbon fibers in a vacuum or H_2/N_2 at 700–1 000°C and 10–25 MPa [42,45,46]. The use of copper-plated carbon fibers alleviates the problem of poor wettability between copper and carbon fibers. On the other hand, alloying additions of Mo, Cr, V, Fe, and Co at 1 at.% to copper improve the wettability [47].

Using continuous unidirectional copper-plated carbon fibers at 35 vol.% in a copper matrix, composites with the following properties parallel to the fibers have been reported [46]:

Thermal conductivity = 270 W/m/K

Thermal expansion coefficient (R.T. to 200°C) = 6×10^{-6}/K

Young's modulus = 150–190 GPa

The thermal conductivity, electrical conductivity and thermal expansion coefficient decrease with increasing fiber volume fraction, while the Young's modulus increases with increasing fiber volume fraction [46]. By using carbon fibers that are graphitic (namely P-120) in the amount of 60 vol.%, a thermal conductivity of 522 W/m/K has been reported [1,48].

Wear resistance is required for brushes. The wear resistance of carbon fiber copper-matrix composites is superior to that of wear-resistant copper alloys [49]. Both the wear rate and the coefficient of friction decrease as the amount of tin in the copper matrix increases, so the use of a Cu–Sn matrix is desirable [50].

Nickel is an attractive matrix because of its heat resistance. However, nickel catalyzes the crystallization of carbon fibers upon heating, resulting in deterioration. Thus, the strength of carbon fiber nickel-matrix composites decreases sharply above 600°C. The Ni-base alloy Hastelloy-D has a higher compatibility with carbon fibers than pure nickel [11]. The activation energy for the C–Ni reaction is 240 kJ/mol, compared to a value of 147 kJ/mol for the C–Al reaction [26].

Carbon fibers have also been used in tin alloy matrices. Short (1 mm long, unplated, 15 vol.%) carbon fiber reinforced Sn–5Sb exhibits a low wear rate and a low coefficient of friction, making it attractive for bearings. The mechanical properties and bearing properties of the composites do not appear to be correlated; unplated (unbonded) carbon fibers give superior bearing properties than copper plated carbon fibers [51]. Continuous copper-plated carbon fibers (29 vol.%) added to a solder alloy (e.g., Sn–40Pb) provide a composite solder material with a low CTE, which is attractive for the bonding of ceramics to ceramics, ceramics to Kovar, etc. The thermal fatigue life (cycling between 25 and 100°C) is increased by up to 87% by the carbon fiber addition. The composite is fabricated by squeeze casting with a mold temperature of 150°C, a pour temperature of 400°C, and a pressure of 30–50 MPa [52].

Carbon fibers have also been used in lead alloy matrices. These composites are made by liquid metal infiltration [53]. They are useful as a positive electrode material in rechargeable lead–acid batteries [54].

Silver-based alloys are widely used as brazing materials. In particular, silver-based titanium-containing alloys, such as Ag–4Ti, are valuable for ceramic brazing because the titanium in the alloy reacts with the ceramic, thereby causing the alloy to wet the ceramic surface. These alloys are called active brazing alloys. Although active brazing alloys are widely accepted for ceramic brazing, they suffer from having high thermal expansion coefficients compared to ceramics. This thermal expansion mismatch between the brazing alloy and the ceramic causes thermal stress. This problem can be alleviated by the addition of carbon fibers to the active brazing alloy, so that the alloy is replaced by a metal-matrix composite. By incorporating about 14 vol.% short copper-coated carbon fibers in Ag–4Ti, the debonding shear strength of brazed joints between stainless steel and alumina was increased by 28%. The use of bare carbon fibers instead of metal-coated carbon fibers is acceptable, because the titanium in the active brazing alloy segregates at the carbon fiber surface, thus helping the fibers to bond to the brazing alloy even without the help of a metal coating. The metal-matrix composite was prepared by adding carbon fibers to a brazing paste. No modification of the brazing operation was necessitated by the carbon fiber addition [55].

Carbon fiber reinforced Ti–Cu (25–35 wt.% Cu) is attractive because titanium has a tendency to wet carbon through the formation of titanium carbide, plus it has a high melting point and a low density. The copper addition serves to retard the titanium carbide formation, as this reaction degrades the fibers. Moreover, copper lowers the melting temperature of titanium. Continuous carbon fiber (8–35 vol.%) Ti–Cu composites are made by heating a mixture of carbon fibers, titanium wires, and copper ribbons close to the liquidus temperature of the alloy to cause infiltration [56].

References

1. C. Thaw, R. Minet, J. Zemany, and C. Zweben, *SAMPE J.* **23**(6), 40–43 (1987).
2. A. Mortensen, J.A. Cornie, and M.C. Flemings, *Materials & Design* **10**(2), 68–76 (1989); *JOM* **40**(2), 12–19 (1988).
3. D.G. Zimcik and B.M. Koike, *SAMPE Q.* **21**(2), 11–16 (1990).
4. J.-M. Chiou and D.D.L. Chung, *J. Mater. Sci.* **28**, 1435–1446 (1993); *J. Mater. Sci.*, **28**, 1447–1470 (1993); *J. Mater. Sci.*, **28**, 1471–1487 (1993).
5. A. Sakamoto, C. Fujiwara, and T. Tsuzuku, *Proc. Jpn. Congr. Mater. Res.*, **33**, 73–79 (1990).
6. I.W. Hall, *Metallography* **20**(2), 237–246 (1987).
7. D.A. Foster, in *Proc. Int. SAMPE Symp. and Exhib.*, *34, Tomorrow's Materials: Today*, edited by G.A. Zakrzewski, D. Mazenko, S.T. Peters, and C.D. Dean, 1989, pp. 1401–1410.
8. M.F. Amateau, *J. Compos. Mater.* **10**, 279 (1976).
9. D.M. Goddard, *J. Mater. Sci.* **13**(9), 1841–1848 (1978).
10. A.P. Levitt and H.E. Band, U.S. Patent 4,157,409 (1979).
11. *Adhesion and Bonding in Composites*, edited by R. Yosomiya, K. Morimoto, A. Nakajima, Y. Ikada, and T. Suzuki, Marcel Dekker, New York, 1990, pp. 235–256. (Chapter on Interfacial Modifications and Bonding of Fiber-Reinforced Metal Composite Material.)
12. D.D. Himbeault, R.A. Varin, and K. Piekarski, in *Proc. Int. Symp. Process. Ceram. Met. Matrix Compos.*, edited by H. Monstaghaci, Pergamon, New York, 1989, pp. 312–323.
13. H. Vincent, C. Vincent, J.P. Scharff, H. Mourichoux, and J. Bouix, *Carbon* **30**(3), 495–505 (1992).
14. H. Katzman, in *Proc. Metal and Ceramic Matrix Composite Processing Conf.*, *Vol. I*, U.S. Dept. of Defense Information Analysis Centers, 1984, pp. 115–140; *J. Mater. Sci.* **22**, 144–148 (1987); *Mater. Manufacturing Processes* **5**(1), 1–15 (1990).
15. L.D. Brown and H.L. Marcus, in *Proc. Metal and Ceramic Matrix Composite Processing Conf.*, *Vol. II*, U.S. Dept. of Defense Information Analysis Centers, 1984, pp. 91–113.
16. G. Leonhardt, E. Kieselstein, H. Podlesak, E. Than, and A. Hofmann, *Mater. Sci. Eng.* **A135**, 157–160 (1991).
17. K. Honjo and A. Shindo, in *Proc. 1st Compos. Interfaces Int. Conf.*, edited by H. Ishida and J. L. Koenig, North-Holland; New York, 1986, pp. 101–107.

18. J.A. Cornie, A.S. Argon, and V. Gupta, *MRS Bull.* **16**(4), 32–38 (1991).
19. H. Landis, Ph.D. dissertation, MIT, 1988.
20. A.S. Argon, V. Gupta, K.S. Landis, and J.A. Cornie, *J. Mater. Sci.* **24**, 1207–1218 (1989).
21. J.P. Clement and H.J. Rack, in *Proc. Am. Soc. Compos. Symp. High Temp. Compos.*, Technomic, Lancaster, PA, 1989, pp. 11–20.
22. S. Schamm, J.P. Rocher, and R. Naslain, in *Proc. 3rd Eur. Conf. Compos. Mater., Dev. Sci. Technol. Compos. Mater.*, edited by A.R. Bunsell, P. Lamicq, and A. Massiah, Elsevier, London, 1989, pp. 157–163.
23. S.N. Patankar, V. Gopinathan, and P. Ramakrishnan, *Scripta Metall.* **24**, 2197–2202 (1990).
24. S.N. Patankar, V. Gopinathan, and P. Ramakrishnan, *J. Mater. Sci. Lett.* **9**, 912–913 (1990).
25. R.V. Subramanian and E.A. Nyberg, *J. Mater. Res.* **7**(3), 677–688 (1992).
26. T. Shinoda, H. Liu, Y. Mishima, and T. Suzuki, *Mater. Sci. Eng.* **A146** (1–2), 91–104 (1991).
27. J.J. Masson, K. Schulte, F. Girot, and Y. Le Petitcorps, *Mater. Sci. Eng.*, **A135**, 59–63 (1991).
28. D. Aidun, P. Martin, and J. Sun, *J. Mater. Eng. Performance* **1**(4), 463–467 (1992).
29. Y. Sawada and M.G. Bader, in *Proc. 5th Int. Conf. Compos. Mater.*, edited by W.C. Harrigan, Jr., J. Shrife, and A.K. Dhingra, Metall. Soc. AIME, Warrendale, PA, 1985, pp. 785–794.
30. R. Wu and W. Cai, in *Proc. 6th Int. Conf. Compos. Mater. 2nd Eur. Conf. Compos. Mater., Vol. 2*, edited by F.L. Matthews, N.C.R. Buskell, J.M. Hodgkinson, and J. Morton, Elsevier, London, 1987, pp. 2.128–2.137.
31. J.J. Masson, K. Schulte, F. Girot, and Y. Le Petitcorps, *Mater. Sci. Eng.* **A135**, 59–63 (1991).
32. Y. Fuwa, H. Michioka and Y. Tatematsu, in *Proc. JSLE Int. Tribol. Conf., Vol. 2*, 1985, pp. 257–262.
33. C. Friend, C. Naish, T.M. O'Brien, and G. Sample, in *Proc. 4th Eur. Conf. Compos. Mater., Dev. Sci. Technol. Compos. Mater.*, edited by J. Fueller, Elsevier, London, 1990, UK, pp. 307–312.
34. D.M. Aylor and R.M. Kain, *ASTM STP 864*, 1985, pp. 632–647.
35. F. Mansfeld and S.L. Jeanjaquet, *Corros. Sci.* **26**(9), 727–734 (1986).
36. X. Dumant, F. Fenot, and G. Regazzoni, in *Proc. Riso Int. Symp. Metallurgy and Materials Science*, 1988, Riso National Lab., Roskilde, Denmark, pp. 349–356.
37. S.S. Tompkins, *ASTM STP 1032*, 1989, pp. 54–67.
38. J.A.G. Furness and T.W. Clyne, *Mater. Sci. Eng.* **A141**(2), 199–207 (1991).
39. R. Chen, R. Wu and G. Zhang, in *Proc. C-MRS Int. Symp., 1990, Vol. 2*, edited by Y. Han, North-Holland, Amsterdam, 1991, pp. 33–37.
40. Y. Wu and G. Zhang, in *Proc. 7th Int. Conf. on Composite Materials, Guangzhou, China, Nov. 1989, Vol. 1*, edited by Y. Wu, Z. Gu, and R. Wu, International Academic Publishers, Beijing, P.R. China, and Pergamon, Oxford, U.K. and New York, 1989, pp. 463–467.
41. K. Kuniya, H. Arakawa, T. Kanai, T. Yasuda, H. Minorikawa, K. Akiyama, and T. Sakaue, in *Proc. 33rd Electron. Compon. Conf.*, 1983, pp. 264–270.
42. D.A. Hutto, J.K. Lucas and W.C. Stevens, in *Proc. Int. SAMPE Symp. and Exhib., 31, Materials Science Future*, 1986, pp. 1145–1153.
43. W. de la Torre, in *Proc. 6th Int. SAMPE Electron. Conf.*, 1992, pp. 720–733.
44. T.W. Chou, A. Kelly, and A. Okura, *Composites* **16**(3), 187–206 (1985).
45. D.N. Fan, Y.L. Wang, H.X. Zhang, Z.N. Liu, and G.J. Li, in *Proc. 7th Int. Conf. on Composite Materials, Guangzhou, China, Nov. 1989, Vol. 1*, edited

by Y. Wu, Z. Gu and R. Wu, International Academic Publishers, Beijing, P.R. China, and Pergamon, Oxford, UK and New York, 1989, pp. 468–474.

46. K. Kuniya and H. Arakawa, in *Composites '86: Recent Advances in Japan and the United States, Proc. Japan–U.S. CCM-III*, edited by K. Kawata, S. Umekawa, and A. Kobayashi, Jpn. Soc. Compos. Mater., Tokyo 1986, pp. 465–472.

47. H. Liu, T. Shinoda, Y. Mishima, and T. Suzuki, *ISIJ International* **29**(7), 568–575 (1989).

48. C. Zweben and K.A. Schmidt, *Electronic Materials Handbook, Vol. 1 (Packaging)*, ASM International, Materials Park, Ohio, 1989, Section 10, Article 10D.

49. K. Kuniya, H. Arakawa, and T. Namekawa, *Trans. Jpn. Inst. Met.* **28**(3), 238–246 (1987).

50. A. Kitamura, T. Teraoka, and R. Sagara, in *Proc. 4th Int. Conf. Compos. Mater., Prog. Sci. Eng. Compos., Vol. 2*, edited by T. Hayashi, 1982, pp. 1473–1480.

51. C.F. Old, I. Barwood and M.G. Nicholas, in *Proc. Inst. Metall.* Spring Meet., *Pract. Met. Compos.*, London, 1974, pp. B47–B50.

52. C.T. Ho and D.D.L. Chung, *J. Mater. Res.* **5**(6), 1266–1270 (1990).

53. C. Wang, M. Ying and D. Yue, in *Proc. 6th Int. Conf. Compos. Mater., 2nd Eur. Conf. Compos. Mater., Vol. 2*, edited by F.L. Matthews, N.C.R. Buskell, J.M. Hodgkinson, and J. Morton, Elsevier, London, 1987, pp. 2.183–2.188.

54. C.M. Dacres, S.M. Reamer, R.A. Sutula and B.F. Larrick, in *Proc. Electrochem. Soc.* **83**(1), 76–95 (1983).

55. M. Zhu and D.D.L. Chung, *J. Am. Ceram. Soc.*, to be published.

56. B. Toloui, in *Proc. 5th Int. Conf. Compos. Mater.*, edited by W.C. Harrigan, Jr., 1985, pp. 773–777.

CHAPTER 8

Carbon-Matrix Composites

Introduction

Carbon fiber carbon-matrix composites, also called carbon–carbon composites, are the most advanced form of carbon, as the carbon fiber reinforcement makes them stronger, tougher, and more resistant to thermal shock than conventional graphite. With the low density of carbon, the specific strength (strength/density), specific modulus (modulus/density) and specific thermal conductivity (thermal conductivity/density) of carbon–carbon composites are the highest among composites. Furthermore, the coefficient of thermal expansion is near zero.

The carbon fibers used for carbon–carbon composites are usually continuous and woven. Both two-dimensional and higher-dimensional weaves are used, though the latter has the advantage of an enhanced interlaminar shear strength. The carbon matrix is derived from a pitch, a resin, or a carbonaceous gas. Depending on the carbonization/graphitization temperature, the resulting carbon matrix can range from being amorphous to being graphitic. The higher the degree of graphitization of the carbon matrix, the greater the oxidation resistance and the thermal conductivity, but the more brittle the material. As the carbon fibers used can be highly graphitic, it is usually the carbon matrix that limits the oxidation resistance of the composite.

The main disadvantages of carbon–carbon composites lie in the high fabrication cost, the poor oxidation resistance, the poor interlaminar properties (especially for two-dimensionally woven fibers), the difficulty of making joints, and the insufficient engineering data base.

Of the world market in carbon–carbon composites, 79% resides in the United States, 20% in Europe and the former USSR, and 1% in Japan. The market is essentially all aerospace, with reentry thermal protection constituting 37%, rocket nozzles constituting 31%, and aircraft brakes constituting 31%. Other applications include furnace heating elements, molten materials transfer, spacecraft and aircraft components, and heat exchangers. Future applications include airbreathing engine components, hypersonic vehicle airframe structures, space structures, and prosthetic devices.

145

Fabrication

The fabrication of carbon–carbon composites is carried out by using four main methods, namely (1) liquid phase impregnation (LPI), (2) hot isostatic pressure impregnation carbonization (HIPIC), (3) hot pressing, and (4) chemical vapor infiltration (CVI).

All of the methods (except, in some cases, CVI) involve firstly the preparation of a prepreg by either wet winding continuous carbon fibers with pitch or resin (e.g., phenolic), or wetting woven carbon fiber fabrics with pitch or resin. Unidirectional carbon fiber tapes are not as commonly used as woven fabrics, because fabric lay-ups tend to result in more interlocking between the plies. For highly directional carbon–carbon composites, fabrics which have a greater number of fibers in the warp direction than the fill direction may be used. After prepreg preparation and, in the case of fabrics, fabric lay-up, the pitch or resin needs to be pyrolyzed or carbonized by heating at 350–850°C. Due to the shrinkage of the pitch or resin during carbonization (which is accompanied by the evolution of volatiles), additional pitch or resin is impregnated in the case of LPI and HIPIC, and carbonization is carried out under pressure in the case of hot pressing. In LPI carbonization and impregnation are carried out as distinct steps, whereas in HIPIC carbonization and impregnation are performed together as a single step.

The carbon yield (or char yield) from carbonization is around 50 wt.% for ordinary pitch and 80–88 wt.% for mesophase pitch [1] at atmospheric pressure. Although mesophase pitch tends to be more viscous than ordinary pitch, making impregnation more difficult, mesophase pitch of viscosity below 1 Pa.sec. at 350°C has been reported [1]. In the case of resins, the carbon yield varies much from one resin to another; for example, it is 57% for phenolic (a thermoset), 79% for polybenzimidazole (PBI, a thermoplast with T_g = 435°C) [2], and 95% for an aromatic diacetylene oligomer [3]. Significant increases in the carbon yield of pitches can be obtained by the use of high pressure during carbonization; at a pressure in excess of 100 MPa, yields as high as 90% have been observed [4]. The higher the pressure, the more coarse and isotropic will be the resulting microstructure, probably due to the suppression of gas formation and escape during carbonization [4]. Spheres, known as mesophase, exhibiting a highly oriented structure similar to liquid crystals and initially around 0.1 μm in diameter, are observed in isotropic liquid pitch above 400°C. Prolonged heating causes the spheres to coalesce, solidify, and form larger regions of lamellar order; this favors graphitization upon subsequent heating to ~ 2 500°C. The high pressure during carbonization lowers the temperature at which mesophase forms [4]. At very high pressures (~ 200 MPa), coalescence of mesophase does not occur. Therefore, an optimum pressure is around 100 MPa [4]. Pressure may or may not be applied during carbonization in LPI but is always applied during carbonization in HIPIC.

In LPI, after carbonization, vacuum impregnation is performed with additional pitch or resin in order to densify the composite. Pressure (e.g., 2

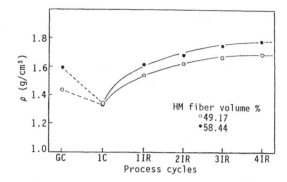

Figure 8.1 Changes of the bulk density of carbon–carbon composites after successive process cycles (GC = green composite; 1C = first carbonization; 1IR, 2IR, 3IR, 4IR = successive impregnation and recarbonization cycles). From Ref. 5. (By permission of Pion, London.)

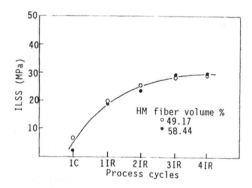

Figure 8.2 Interlaminar shear strength (ILSS) of carbon–carbon composites after successive impregnation/recarbonization cycles. Notation as in Fig. 8.1. From Ref. 5. (By permission of Pion, London.)

MPa) may be applied to help the impregnation. The carbonization–impregnation cycles are repeated several times (typically 3–6) in order to achieve sufficient densification. Figures 8.1 and 8.2 show the effects of the number of carbonization–impregnation cycles on the density and interlaminar shear strength (ILSS), respectively, of carbon–carbon composites prepared from pitch and PAN-based HM-type carbon fibers (Torayca M40B) [5]. Both the density and ILSS increase with increasing number of cycles. Figure 8.1 also shows that the first carbonization (1C) decreases the density from the value of the green composite (GC), so that subsequent impregnation and recarbonization are necessary [5].

As shown in Figure 8.1, the density levels off after a few cycles of impregnation and recarbonization. This is because the repeated densification cycles cause the mouths of the pores to narrow down, so that it is difficult for the impregnant to enter the pores. As a consequence, impregnant pickup levels

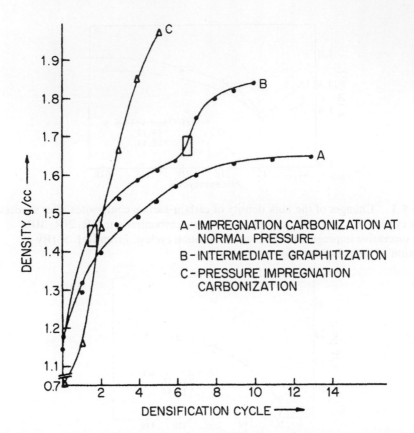

Figure 8.3 Variation of the density of carbon–carbon composites with impregnation cycle by three processes. Process C is HIPIC. Processes A and B are the same except that A has no intermediate graphitization whereas B does (at each of the two rectangles along the curve for B). From Ref. 7. (By permission of Publications & Information Directorate, India.)

off. This problem can be alleviated by intermediate graphitization, wherein the composites are subjected to a heat treatment at 2 200–3 000°C between the carbonization and impregnation steps after the densification cycle when the density levels off. On graphitization, the entrance to the pores opens up due to rearrangement of the crystallites in the matrix. These opened pores then become accessible during further impregnation, thus leading to further density increase [6,7]. Figure 8.3 [7] shows the effect of two 2 700°C graphitizations carried out at the intermediate stages of fabrication when a saturation in the density increase is observed. The intermediate graphitizations allow the density to reach 1.84 g/cm³ (B in Figure 8.3), compared to a density of 1.65 g/cm³ (A in Figure 8.3) for the case of impregnation carbonization at a normal pressure (2 MPa) without intermediate graphitization. The tensile strength was increased by up to 74% by the addition of a 2 600°C graphitization step to just one of the cycles [6]. The fracture toughness was increased one-fold by

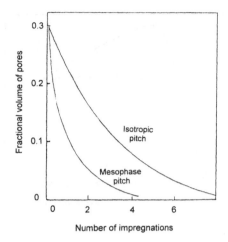

Figure 8.4 Plot of residual porosity versus number of impregnation cycles for mesophase and isotropic pitches. From Ref. 9. (By permission of IOP Publishing Limited.)

intermediate graphitization [8]. The use of mesophase pitch instead of isotropic pitch for impregnation can cut down on the required number of impregnation cycles, as shown in Figure 8.4 [9].

In HIPIC, an isostatic inert gas pressure of around 100 MPa is applied to impregnate pitch (rather than resins, which suffer from a low carbon yield) into the pores in the sample while the sample is being carbonized at 650–1 000°C. The pressure increases the carbon yield and maintains the more volatile fractions of the pitch in a condensed phase. After this combined step of carbonization and impregnation, graphitization is performed by heating without applied pressure above 2 200°C. Figure 8.3C [7] shows that HIPIC allows the density to reach a higher value than LPI (with or without intermediate graphitization) and that a smaller number of cycles is needed for HIPIC to achieve the high density. However, HIPIC is an expensive technique.

One HIPIC process involves vacuum impregnating a dry fiber preform or porous carbon–carbon laminate with molten pitch, placing it inside a metal container (or can) with an excess of pitch surrounding it inside the can. The can is then evacuated and sealed (preferably by using an electron beam weld) and placed within the work zone of a hot isostatic press (HIP) unit. The temperature is then raised at a programmed rate above the melting point of the pitch, but not so high as to result in weight loss due to the onset of carbonization. The pressure is then increased and maintained at around 100 MPa. The pitch initially melts and expands within the can and is forced by isostatic pressure into the pores in the sample. The sealed container acts like a rubber bag, facilitating the transfer of pressure to the workpiece. After that, the temperature is gradually increased towards that required for pitch carbonization (650–1 000°C). The pressure not only increases the carbon yield,

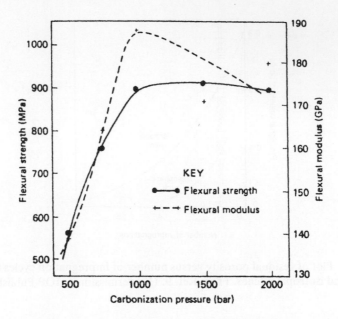

Figure 8.5 Effect of carbonization pressure on the mechanical properties of carbon–carbon composite prepared by HIPIC. From Ref. 10. (By permission of the publishers, Butterworth–Heinemann Ltd.)

but also prevents liquid from being forced out of the pores by pyrolysis products. After the HIPIC cycle is complete, the preforms are removed from their container and cleaned up by removing any excess carbonized liquid from the surface [10]. Figure 8.5 [10] shows that the optimum carbonization pressure for HIPIC is 1 000–1 500 bar (100–150 MPa). Lower pressures are insufficient to prevent bloating of the composites due to the evolution of carbonization gases. Higher pressures do not offer any significant improvement, and even seem to be detrimental to the mechanical properties [10].

Another HIPIC process does not use a can, but simply applies an isostatic gas pressure on the surface of molten pitch, which seals the workpiece as illustrated in Figure 8.6 [11]. HIPIC increases the carbon yield of pitch, especially when the molecular size of the pitch is small. Table 8.1 [11] shows the carbon yield at 0.1 and 10 MPa for pitches of three different molecular weights. The increase in pressure causes the carbon yield to increase dramatically for pitch A (low molecular weight), but only slightly for pitch C (high molecular weight). This is due to the already high carbon yield of pitch C at 0.1 MPa. Table 8.2 [11] shows the basic characteristics of pitches A, B, and C. Pitch C has a high fixed carbon content, a high content of insoluble quinoline and a high softening point. The improvement in carbon yield due to the use of pressure becomes saturated at a pressure of 10 MPa. The origin of the improvement is attributed to the trapping and decomposition of the evolved hydrocarbon gases under high pressure; the decomposition produces

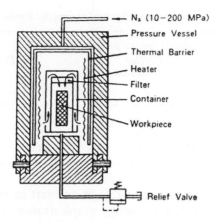

Figure 8.6 HIPIC (pressure carbonizing) furnace. From Ref. 11. (Reprinted by courtesy of Elsevier Sequoia S.A., Lausanne, Switzerland.)

Table 8.1 Pitch properties. From Ref. 11.

Pitch	Aromatization ratio	Molecular weight	Carbon yield (%) 0.1 MPa	Carbon yield (%) 10 MPa
A	1.063	726	45.2	85.9
B	1.305	782	54.4	86.4
C	0.618	931	84.5	89.8

Table 8.2 Matrix precursor. From Ref. 11.

Pitch	Type	Fixed C (%)	Quinoline (%)	Toluene (%)	Softening point (°C)
A	Coal tar	52	20	0.3	83
B	Petroleum	54	19.4	1.2	131
C	Petroleum	82	37	0	266

carbon and hydrogen. Table 8.3 [11] shows the effect of pressure on the bulk density, porosity, and flexural strength. The increase in pressure causes the bulk density to increase, the porosity to decrease and the flexural strength to increase. Moreover, as the pressure rises, the pores in the carbonized matrices become smaller in size and more spherical in shape [11].

In hot pressing (also called high-temperature consolidation), carbonization is performed at an elevated temperature (1 000°C typically, but only 650°C for an aromatic diacetylene oligomer as the matrix precursor [3]) under a uniaxial pressure (2–3 MPa typically, but 38–76 MPa for an aromatic diacetyl-

Table 8.3 Bulk density, porosity, and flexural strength. From Ref. 11.

Pressure (MPa)	Bulk density (g/cm³) for the following number of cycles					Porosity (%)	Flexural strength (kgf/mm²)
	1	2	3	4	7		
0.1	1.08	1.22	1.44	–	1.73	14	45
10	1.24	1.34	1.49	1.75	1.75	12	53
200	1.34	1.57	1.98	1.98	1.98	6	60

ene oligomer as the matrix precursor [3]) in an inert or reducing atmosphere, or in a vacuum. During hot pressing, graphitization may occur even for thermosetting resins, which are harder to graphitize than pitch [12]. This is known as stress-graphitization. Subsequently, further graphitization may be performed by heating without applied pressure at 2 200–3 000°C. No impregnation is performed after the carbonization. Composites made by hot pressing have flattened pores in the carbon matrix and the part thickness is reduced by about 50%. Excessive pressure (say, 5 MPa) causes the formation of vertical cracks [13].

In CVI (also called CVD, chemical vapor deposition), gas phase impregnation of a hydrocarbon gas (e.g., methane, propylene) into a carbon fiber preform takes place at 700–2 000°C, so that pyrolytic carbon produced by the cracking of the gas is deposited in the open pores and surface of the preform. The carbon fiber preform can be in the form of carbon fabric prepregs which have been carbonized and graphitized, or in the form of dry wound carbon fibers. There are three CVI methods, namely the isothermal method, the temperature gradient method, and the pressure gradient method.

In the isothermal method, the gas and sample are kept at a uniform temperature. As carbon growth in the pores will cease when they become blocked, there is a tendency for preferential deposition on the exterior surfaces of the sample. This causes the need for multiple infiltration cycles, such that the sample is either skinned by light machining or exposed to high temperatures to reopen the surface pores for more infiltration in subsequent cycles.

In the temperature gradient method, an induction furnace is used. The sample is supported by an inductively heated mandrel (a susceptor) so that the inside surface of the sample will be at a higher temperature than the outside surface. The hydrocarbon gas flows along the outside surface of the sample. Due to the temperature gradient, the deposition occurs first at the inside surface of the sample and progresses toward the outside surface, thereby avoiding the crusting problem.

In the pressure gradient method, the hydrocarbon gas impinges on to the inside surface of the sample, so the gas pressure is higher at the inside surface than the outside surface. The pressure gradient method is not as widely used as the isothermal method or the temperature gradient method.

Both the temperature gradient method and the pressure gradient method are limited to single samples, whereas the isothermal method can handle several samples at once. However, the isothermal method is limited to thin samples due to the crusting problem.

A drawback of CVI is the low rate of deposition resulting from the use of a low gas pressure (1–150 torr), which favors a long mean free path for the reactant and decomposed gases; a long mean free path enhances deposition into the center of the sample. A diluent gas (e.g. He, Ar) is usually used to help the infiltration. For example, a gas mixture containing 3–10 vol.% propylene (C_3H_6) in Ar was used for CVI at 760–800°C [14]. Hydrogen is often used as a carbon surface detergent.

An attraction of CVI is that CVI carbon is harder than char carbon from pitch or resin, so that CVI carbon is particularly desirable for carbon–carbon composites used for brakes and friction products.

Fillers such as carbon black can be added to the resin or pitch prior to carbonization in order to provide bridging between the fibers during subsequent CVI [15].

The methods of CVI and LPI may be combined by first performing CVI on dry wound carbon fibers then performing LPI [16]. The CVI step serves to make the carbon fibers rigid prior to impregnation.

The quality of a carbon–carbon composite depends on the quality of the polymer-matrix composite from which the carbon–carbon composite is made. For example, resin pooling may result in areas of excessive shrinkage cracks after carbonization and graphitization at high temperatures [15].

During carbonization, pitch tends to bloat due to the evolution of gases generated by pyrolysis. This can cause the expulsion of pitch from the carbon fiber preform during carbonization. There are two ways to alleviate this problem. One way is oxidation stabilization, which is oxidation of the pitch at a temperature below the softening point of the pitch, i.e. generally below 300°C [17]. Another way is uniaxial pressing at 500 Pa and between room temperature and 600°C prior to carbonization at 1 000°C [18].

The choice of pitch affects the carbon yield, which increases with increasing C/H ratio of the pitch [19]. Mesophase pitch gives a higher carbon yield than isotropic pitch; it can be prepared by removal of the light aromatic materials by solvent extraction with toluene [18].

The combined use of pitch and resin is also possible. Resin (epoxy) can be used to form carbon fiber prepreg sheets, which are then laminated alternately in a die with pitch, which is in the form of a mixture of coal tar pitch, coke powder, and carbonaceous bulk mesophase (as binder for the coke). Carbon–carbon composites were thus prepared by hot pressing at 600°C and 49 MPa, followed by heat treatment in N_2 at 1 500°C [20].

The choice of carbon fibers plays an important role in affecting the quality of the carbon–carbon composites.

The use of graphitized fibers (fired at 2 200–3 000°C) with a carbon content in excess of 99% is preferred, because their thermal stability reduces the part warpage during later high temperature processing to form a carbon–carbon composite [15]. Moreover, graphitized fibers lead to better densified carbon–carbon composites than carbon fibers which have not been graphitized [21]. This is because the adhesion of the fibers with the polymer resin is weaker for graphitized fibers, so that, on carbonization, the matrix can easily shrink away from the fibers, leaving a gap which can be filled in during subsequent impregnation [21]. In contrast, the polar surface groups on carbonized fibers make strong bonds with the resin (phenolic), thus inhibiting the shrinking away of the charred matrix from the fibers and leading to the formation of fine microcracks in the carbon matrix [21].

Circular fibers are preferred to irregularly shaped fibers, as the latter lead to stress concentration points in the matrix around the fiber corners. Microcrack initiation occurs at these points, thus resulting in low strength in the carbon–carbon composite [22].

The microstructure of mesophase pitch-based carbon fibers influences the physical changes that take place during graphitization of the carbon–carbon composite made with these fibers. For medium-modulus carbon fibers having parallel graphite planar sheetlike microstructure, the prestressed carbon matrix shears and orders itself in the fiber direction during graphitization, thereby stretching the fibers. This causes an expansion of the composite in the fiber direction and an increase in the flexural strength of the composite after graphitization. In contrast, for mesophase pitch-based carbon fibers having sheath- and core-type microstructures, the composite does not expand in the fiber direction and the flexural strength of the composite decreases upon graphitization [23].

The weave pattern of the carbon fabric affects densification. The 8H satin weave is preferred over plain weave because of the inhomogeneous matrix distribution around the crossed bundles in the plain weave. Microcracks develop beneath the bundle crossover points. After carbonization, composites made with the plain weave fabric show nearly catastrophic failure with bundle pullout, whereas those made with the satin weave fabric show shear-type failure with fiber pullout. On densification, the flexural strength of the composites made with the satin weave fabric increases appreciably, whereas only marginal improvement is obtained in composites made with the plain weave fabric [21].

The fiber–matrix bond strength in carbon–carbon composites must be optimum. If the bond strength is too high, the resulting composite may be extremely brittle, exhibiting catastrophic failure and poor strength. If it is too low, the composites fail in pure shear, with poor fiber strength translation [24]. Thus, among (1) nonsurface-treated unsized carbon fibers (too low in bond strength), (2) nonsurface-treated sized carbon fibers (optimum), (3) surface-treated unsized carbon fibers (too high in bond strength) and (4) surface-treated sized carbon fibers (too high in bond strength), the nonsurface-treated sized carbon fibers give carbon–carbon composites of the highest strength. These optimum composites fail in a mixed mode fracture [24]. Similarly,

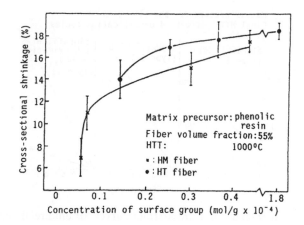

Figure 8.7 Composite cross-sectional shrinkage after carbonization at 1 000°C versus concentration of reactive surface groups for composites with untreated and surface-treated (conc. HNO_3) fibers of types HT and HM. From Ref. 26. (Reprinted by permission of Kluwer Academic Publishers.)

carbon fibers that have been oxygen plasma treated and then argon plasma detreated (optimum) give carbon–carbon composites of higher strength than carbon fibers that have been oxygen plasma treated (but not argon plasma detreated) (too high in bond strength) and than those which have not been treated at all (too low in bond strength) [25]. Moreover, carbon fibers which have been oxidized by nitric acid and then detreated in an argon plasma give composites of higher strength than those which have been oxidized in nitric acid but not detreated [25].

Surface treatment (say, with concentrated nitric acid) of carbon fibers increases the concentration of surface groups, thus strengthening the fiber–matrix bonding. The strengthened fiber–matrix bonding makes it more difficult for the matrix to shrink away from the fiber surface during carbonization, so the fibers get pulled together by the matrix shrinkage and a large composite cross-sectional shrinkage results. The relationship between the composite cross-sectional shrinkage and the surface group concentration is depicted in Figure 8.7, which is for a fiber volume fraction of 55%, a carbonization temperature of 1 000°C, and phenolic resin as the matrix precursor [26].

Figure 8.7 also shows that HT fiber composites yield greater shrinkage than HM fiber composites [26]. This is due to the greater density of active sites in HT fibers [27] and the resulting stronger fiber–matrix bonding for HT fibers. The greater density of active sites in HT fibers is due to the lower heat treatment temperature used in the fabrication of HT fibers compared with HM fibers, as surface defects tend to be reduced by annealing. The correlation between surface group concentration and composite cross-sectional shrinkage has been demonstrated for four types of fibers [28].

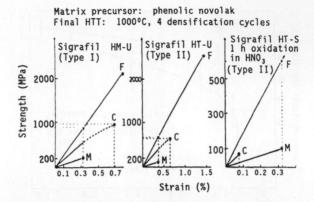

Figure 8.8   Experimental flexural stress–strain curves for unidirectional carbon–carbon composites after carbonization at 1 000°C. Left: untreated HM fibers; middle: untreated HT fibers; right: surface treated HT fibers. F = fiber; C = composite; M = matrix. From Ref. 26. (Reprinted by permission of Kluwer Academic Publishers.)

Figure 8.8 shows experimental flexural stress–strain curves of 1 000°C carbonized composites of untreated HM fibers, untreated HT fibers, and surface-treated HT fibers. The composite with untreated HM fibers is one with the weakest fiber–matrix bonding and it is the one with the highest strength and the highest strain at failure [26]. Thus, excessive fiber–matrix bonding is detrimental to the mechanical properties of the carbonized composites.

A fiber surface treatment that is too strong for a carbon–carbon composite which has not been graphitized may be optimum when the composite has been graphitized. This is because graphitization reduces the fiber–matrix bond strength. Thus, comparison of composites made with fibers that have been treated with nitric acid for 0, 15, 60, and 300 min. shows that the 15 min. treatment is optimum for composites before graphitization, but a treatment ⩾ 300 min. is optimum for composites after graphitization [29]. For composites comprising surface-treated fibers, the flexural strength of the composites heat-treated at 2 000°C is higher than that of the composites treated at 1 000°C; for the composites with untreated fibers, the results are opposite [30].

The graphitization of a carbon–carbon composite is significantly greater than that of fiber or matrix alone to the same temperature. The graphitization of the composite is mainly confined to the matrix, usually as a sheathlike structure adjacent to the fiber and 1–3 μm thick [31]. The sheaths can be observed from the composite's fracture surface (Figure 8.9 [32]) and etched metallographic section (Figure 8.10 [33]). Such localized graphitization, termed stress graphitization, is believed to be the result of thermally induced tensile or compressive stresses acting at the fiber–matrix interface. These thermal expansion stresses vary with different fiber–matrix combinations. Debonded regions show less stress graphitization than well-bonded regions, because the debond gaps impede stress buildup at the fiber–matrix interface [31].

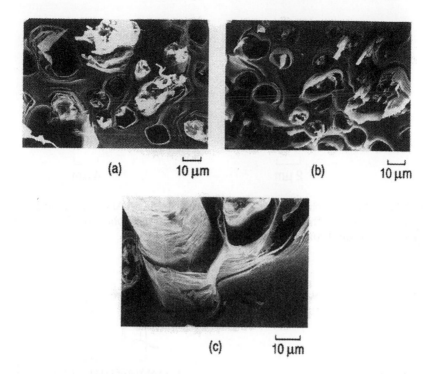

Figure 8.9 Fracture surface of a carbon–carbon composite heat-treated at 2 750°C, showing matrix sheath tubes. From Ref. 32.

Polyarylacetylene (PAA) is a resin that is typically nongraphitizing, but it is attractive in its high char yield at 88% (compared to a char yield of ~50% for the phenolic resin) and it has easy processibility compared to pitch [34]. The chemical structure and processing of PAA-based composites are illustrated in Figure 8.11 [35]. PAA is derived from the polymerization of diethynylbenzene (monomer) [35]. PAA can be catalytically transformed to a graphitic-type carbon by the addition of boron in the form of a carborane compound, $C_2B_{10}H_{12}$. The extent of graphitization is controlled by the amount of catalyst present and the heat treatment temperature. The onset of catalytic graphitization occurs at temperatures much lower than typically used in carbon–carbon composite processing [34].

Oxidation Protection

In the absence of oxygen, carbon–carbon composites have excellent high-temperature strength, modulus, and creep resistance. For example, the carbon–carbon composites used for the nose cap of the Space Shuttle can withstand 1 600°C, whereas more advanced carbon–carbon composites can

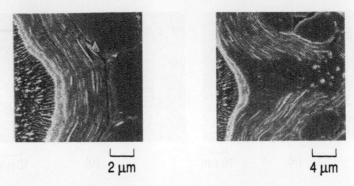

2 μm   4 μm

Figure 8.10 An ion-etched section of a carbon–carbon (PAA/T-50) composite heat-treated to 2 900°C. The striated region to the left of each photograph is etched epoxy mounting resin. From Ref. 33.

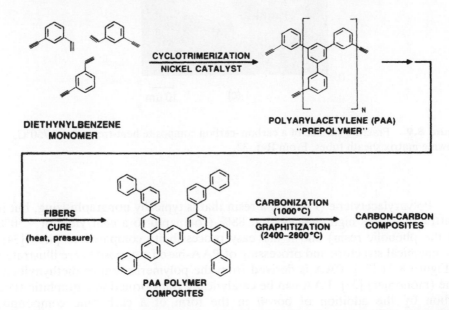

Figure 8.11 Chemical structure and processing of PAA-based composites. From Ref. 35.

withstand 2 200°C. In contrast, superalloys can withstand only 1 200°C and also suffer from having high densities.

Carbon–carbon composites are highly susceptible to oxidation at temperatures above 320°C [36]. The predominant reaction that occurs in air is:

$$C + O_2 \rightarrow CO_2 \uparrow$$

This reaction is associated with a very large negative value of the Gibbs free energy change, so it proceeds with a big driving force even at very low O_2

partial pressures [37]. Thus the rate of oxidation is controlled not by the chemical reaction itself, but by transport of the gaseous species to and away from the reaction front [37].

The oxidation of carbon–carbon composites preferentially attacks the fiber–matrix interfaces and weakens the fiber bundles. The unoxidized material fails catastrophically by delamination cracking between plies and at bundle–bundle interfaces within plies. As oxidation progresses, failure becomes a multistep process with less delamination cracking and more cross-bundle cracking. This change of failure mode with oxidation is attributed to more severe attack within bundles than at bundle–bundle interfaces. For a weight loss on oxidation of 10%, the reductions in elastic modulus and flexural strength were 30% and 50%, respectively [38].

Oxidation protection of carbon–carbon composites up to 1 700°C involves various combinations of four methods.

1. SiC coatings applied by pack cementation, reaction sintering, silicone resin impregnation/pyrolysis, or chemical vapor deposition (CVD) to the outer surface of the composite.
2. Oxidation inhibitors (oxygen getters and glass formers) introduced into the carbon matrix during lay-up and densification cycles to provide additional oxidation protection from within by migrating to the outer surface and sealing cracks and voids during oxidation.
3. Application of a glassy sealant on top of the SiC conversion coating mainly by slurry brush-on, so that the sealants melt, fill voids and stop oxygen diffusion, and, in some cases, act as oxygen getters.
4. Dense SiC or Si_3N_4 overlayers applied by chemical vapor deposition (CVD) on top of the glassy sealant, if a glassy sealant is used, or, otherwise, on top of the SiC conversion coating, to control and inhibit the transfer of oxygen to the substrate and to control venting of reaction products to the outside [36,39].

A SiC coating in Method 1, known as SiC conversion coating, is gradated in composition so that it shades off gradually from pure silicon compounds on the outside surface to pure carbon on the inside [40]. The gradation minimizes spallation resulting from the thermal expansion mismatch between SiC and the carbon–carbon composite. The conversion coating can also be made to be gradated in porosity so that it is denser near the outside surface [36]. The SiC conversion coating is applied by pack cementation, which involves packing the composite in a mixture of silicon carbide and silicon powders then heating this assembly up to 1 600°C. During this process, primarily the following reactions take place [29]:

$$Si(l) + C \rightarrow SiC$$

$$Si(g) + C \rightarrow SiC$$

$$SiO(g) + 2C \rightarrow SiC + CO(g)$$

The net result is the chemical conversion of the outermost surfaces of the composite to silicon carbide. Typical thicknesses of pack cementation coatings range from 0.3 to 0.7 mm [36]. One disadvantage of this process is that elemental silicon may be trapped in the carbon matrix under the conversion coating. The entrapped silicon tends to vaporize at elevated temperatures and erupt through the coating, leaving pathways for oxygen to migrate to the carbon–carbon substrate [36].

A second method to form a SiC coating is reaction sintering, which involves dipping a carbon–carbon composite into a suspension of silicon powder (average 10 μm size) in an alcohol solution then sintering at 1 600°C for 4 h. in argon [41].

A third method to form a SiC coating involves vacuum impregnating and cold isostatic pressing (30 000 psi or 200 MPa) a silicone resin into the matrix of a carbon–carbon composite and subsequent pyrolysis at 1 600°C for 2 h. in argon [41].

The SiC overlayer in Method 4 is more dense than the SiC conversion coating in Method 1. It serves as the primary oxygen barrier [39]. It is prepared by CVD, which involves the thermal decomposition/reduction of a volatile silicon compound (e.g., CH_3SiCl, CH_3SiCl_3) to SiC. The reaction is of the form [42]:

$$CH_3SiCl_3(g) \xrightarrow{\text{heat/}H_2} SiC + 3HCl(g)$$

The decomposition occurs in the presence of hydrogen and heat (e.g., 1 125°C [43]). If desired, the overlayer can be deposited so that it contains a small percentage of unreacted silicon homogeneously dispersed in the SiC [29]. The excess silicon upon oxidation becomes SiO_2, which has a very low oxygen diffusion coefficient. Such silicon-rich SiC is abbreviated SiSiC. Instead of SiC, Si_3N_4 may be used as the overlayer; Si_3N_4 can also be deposited by CVD.

Silicon-based ceramics (SiC and Si_3N_4), among high-temperature ceramics, have the best thermal expansion compatibility with respect to carbon–carbon composites and exhibit the lowest oxidation rates. Moreover, the thin amorphous SiO_2 scales that grow have low oxygen diffusion coefficients and can be modified with other oxides to control the viscosity [44]. Above 1 800°C, these silicon-based ceramic coatings cannot be used because of the reactions at the interface between the SiO_2 scale and the underlying ceramic. Furthermore, the reduction of SiO_2 by carbon produces CO(g) [44].

The glass sealants in Method 3 are in the form of glazes comprising usually silicates and borates. If desired, the glaze can be filled with SiC particles [44]. The sealant is particularly important if the SiC conversion coating is porous. Moreover, when microcracks develop in the dense overlayer, the sealant fills the microcracks. Borate (B_2O_3) glazes wet C and SiC quite well, but they are useful up to 1 200°C due to volatilization [45,46]. Moreover, B_2O_3 has poor moisture resistance at ambient temperatures, as it undergoes hydrolysis, which results in swelling and crumbling [44]. In addition, B_2O_3 has a tendency to galvanically corrode SiC coatings at high temperatures [47]. However, these problems of B_2O_3 can be alleviated by the use of multicomponent systems, such as $10TiO_2 \cdot 20SiO_2 \cdot 70B_2O_3$ [47]. TiO_2 has a high solubility in B_2O_3; it is used to prevent volatilization of B_2O_3 and increase the viscosity over a wide temperature range. The SiO_2 component acts to increase the moisture resistance at ambient temperatures, reduce the B_2O_3 volatility at high temperatures, increase the overall viscosity of the sealant, and prevent galvanic corrosion of the SiC at high temperatures by the B_2O_3 [47].

The inhibitors in Method 2 are added to the carbon matrix by incorporation as particulate fillers in the resin or pitch prior to prepregging. They function as oxygen getters and glass formers. These fillers can be in the form of elemental silicon, titanium, and boron. Oxidation of these particles within the carbon matrix forms a viscous glass, which serves as a sealant that flows into the microcracks of the SiC coating, covering the normally exposed carbon–carbon surface to prevent oxygen ingress into the carbon–carbon [47]. The mechanism of oxidation inhibition by boron-based inhibitors may involve B_2O_2, a volatile suboxide that condenses to B_2O_3 upon encountering a locally high oxygen partial pressure in coating cracks [48]. Instead of using elemental Ti and Si, a combination of SiC, Ti_5Si_3, and TiB_2 may be used [49]. For a more uniform distribution of the glass sealant, the filler ingredients may be prereacted to form alloys such as Si_2TiB_{14} prior to addition to the resin or pitch [49]. Yet another way to obtain the sealant is to use an organoborosilazane polymer solution [50].

The addition of glass-forming additives, such as boron, silicon carbide and zirconium boride, to the carbon matrix can markedly reduce the reactivity of the composite with air, but the spreading of the glassy phase throughout the composite is slow and substantial fractions of the composite are gasified before the inhibitors become effective. Thus, in the absence of an exterior impermeable coating, the oxidation protection afforded at temperatures above $1\,000°C$ by inhibitor particles added to the carbon matrix is strictly limited [51].

The inhibition mechanism of B_2O_3 involves the blockage of active sites (such as the edge carbon atoms) for small inhibitor contents and the formation of a mobile diffusion barrier for oxygen when the B_2O_3 amount is increased [52,53]. Figure 8.12 [53] shows the effect of B_2O_3 addition (in amounts of 3 and 7 wt.%) on the oxidation resistance in air at 710°C. The inhibition effect of B_2O_3 is most pronounced at the beginning of oxidation, as shown by the small slope of the weight loss curves near time zero. Thereafter a pseudolinear oxidation regime takes place, as for the untreated composite. The inhibition factor is defined as the ratio of the oxidation rate of the untreated carbon to that of the treated carbon. Figure 8.13 [53] shows the dependence of the inhibition factor on the B_2O_3 content for a temperature of 710°C and a burn-off level of 20 wt.%. For a B_2O_3 content below 2 wt.%, the inhibition factor increases sharply with the B_2O_3 content. Thereafter a more gradual linear increase takes place.

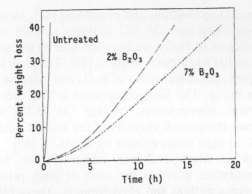

Figure 8.12 Carbon weight loss upon oxidation in air at 710°C in the presence (2 or 7 wt. %) or absence (untreated) of B_2O_3. From Ref. 51. (Reprinted with permission from Pergamon Press Ltd.)

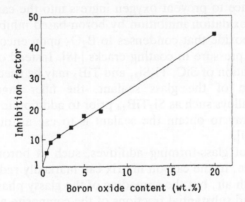

Figure 8.13 The inhibition factor as a function of the B_2O_3 content. From Ref. 51. (Reprinted with permission from Pergamon Press Ltd.)

For oxidation protection above 1 700°C, a four-layer coating scheme is available. This scheme consists of a refractory oxide (e.g. ZrO_2, HfO_2, Y_2O_3, ThO_2) as the outer layer for erosion protection, an SiO_2 glass inner layer as an oxygen barrier and sealant for cracks in the outer coating, followed by another refractory oxide layer for isolation of the SiO_2 from the carbide layer underneath, and finally a refractory carbide layer (e.g., TaC, TiC, HfC, ZrC) to interface with the carbon–carbon substrate and to provide a carbon diffusion barrier for the oxide to prevent carbothermic reduction. The four-layer system is thus refractory oxide/SiO_2 glass/refractory oxide/refractory carbide [4,44]. It should be noted that ZrO_2, HfO_2, Y_2O_3 and ThO_2 have the required thermal stability for long-term use at $> 2\,000$°C, but they have very high oxygen permeabilities; silica exhibits the lowest oxygen permeability and is the best candidate as an oxygen barrier other than iridium at $> 1\,800$°C; iridium suffers from a relatively high thermal expansion coefficient, high cost, and limited availability [44].

A ternary HfC–SiC–HfSi$_2$ system deposited by CVD has been reported to provide good oxidation protection up to 1 900°C [50]. The HfC component is chemically compatible with carbon. Furthermore, HfO$_2$ forms from HfC by the reaction:

$$HfC + \frac{3}{2} O_2 \rightarrow HfO_2 + CO$$

and HfO$_2$ is a very stable oxide at high temperatures. However, HfO$_2$ undergoes a phase change from monoclinic to tetragonal at 1 700°C, with a volume change of 3.4%. To avoid catastrophic failure due to the volume change, HfO$_2$ is stabilized through the addition of HfSi$_2$. The SiC component acts as a diffusion barrier [54].

Pack cementation is a relatively inexpensive technique for coating carbon–carbon composites in large quantities. The quality of SiC coatings prepared by pack cementation can be improved by first depositing a 10 μm carbon film by CVD on to the surface of the carbon–carbon composite, because the carbon film improves the homogeneity of the carbon–carbon surface and eases the reaction with Si [55]. Similarly carbon CVD can be used to improve SiC films deposited by reaction sintering or resin impregnation [41]. The carbon CVD involves the pyrolysis of methane in a tube furnace at 1 300°C [41].

Pack cementation has been used to form chromium carbide coatings in addition to SiC coatings for oxidation protection of carbon–carbon composites. For chromium carbide coatings, the carbon–carbon composite sample is packed in a mixture of chromium powder, alumina powder, and a small quantity of NH$_4$Cl (an activator) and reacted at 1 000°C in argon. The chromium powder produces chromium carbide by reaction with the carbon–carbon composite. At the same time, HCl dissociated from NH$_4$Cl reacts with the chromium powder to form chromous halide liquid, which reacts with the carbon–carbon composite to form chromium carbide. The latter kind of chromium carbide permeates the openings in the former kind of chromium carbide. Upon oxidation of the chromium carbide coating, a dense layer of Cr$_2$O$_3$ is formed and serves to prevent oxidation of the carbon–carbon composite [56].

Another form of oxidation protection can be provided by treatments of carbon–carbon composites by various acids [57] and bromine [58].

The fundamental approaches for oxidation protection of carbons can be categorized into four groups [59]: (1) prevention of catalysis, (2) retardation of gas access to the carbon, (3) inhibition of the carbon-gas reactions, and (4) improvement in the carbon crystallite structure. Approach 2 is the dominant one applied to carbon–carbon composites, as it provides the greatest degree of oxidation protection. However, the other approaches need to be exploited as well. In particular, Approach 4 means that pitch and CVI carbon are preferred to resins as precursors for carbon–carbon matrices, as pitch and CVI carbon are more graphitizable than resins [60]. Nevertheless, the stress applied to the matrix by the adjacent fibers during carbonization causes alignment of the matrix molecules near the fibers [61]. Furthermore, the microstructure of the carbon fibers affect strongly the microstructure of the carbon matrix, even when the fiber fraction is only about 50 wt.%, and the microstructure of the carbon matrix affects the amount of accessible porosity, thereby affecting the oxidation behavior [62].

The application of coatings on carbon–carbon composites can deteriorate the room temperature mechanical properties of carbon–carbon composites. For example, after application of a 0.25–0.50 mm thick SiC conversion coating, the flexural strength decreases by 29% [36]. On the other hand, oxidation of a carbon–carbon composite to a burn-off of 20% causes the flexural strength to decrease by 64% [63].

Mechanical Properties

Figures 8.14 and 8.15 show the flexural strength and flexural modulus of carbon–carbon composites containing continuous HM fibers as functions of the composite density. Both flexural strength and modulus increase with increasing density for a constant fiber volume fraction of either 49.17 or 58.44%. The flexural strength increases linearly with the density but levels out at around 500 MPa (Figure 8.14). The flexural modulus increases exponentially with increasing density (Figure 8.15); its value is more than twice that in composites with resin matrix [5].

The mechanical properties of carbon–carbon composites are much superior to those of conventional graphite. Three-dimensional carbon–carbon composites are particularly attractive. Their preform structure can be tailored in three directions. The three-dimensional integrated preform structure results in superior damage tolerance and minimum delamination crack growth under interlaminar shearing compared with two-dimensional laminate carbon–carbon composites. Unlike conventional materials, the crack in three-dimensional carbon–carbon composites diffuses in a tortuous manner, probably tracking preexisting voids or microcracks. The failure of three-dimensional composites involves a series of stable crack propagation steps across the matrix and yarn bundles, followed by unstable crack propagation. The dominant damage mechanisms are bundle breakage and matrix cracking [64].

Table 8.4 compares the properties of three carbon–carbon composites (labeled A, B, and C) and an isostatically molded fine-grain petroleum coke graphite (labeled G) [65]. Composites A and B use carbon fibers in the form of felt. The felt in A is based on pitch and constitutes 47 wt.% of composite A; the felt in B is based on PAN and constitutes 34 wt.% of composite B. Composite C is a two-dimensional composite with carbon fibers in the form of rayon-based carbon fiber cloths stacked on top of one another. The heat treatment temperature is 3 000°C for all these samples (A, B, C, and G). Table

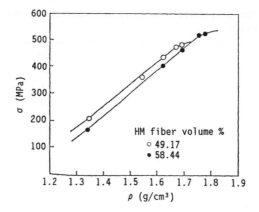

Figure 8.14 Dependence of the flexural strength (σ) on the bulk density (ρ) of carbon–carbon composites carbonized at 1 000°C. From Ref. 5. (By permission of Pion, London.)

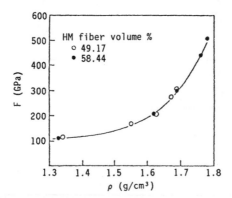

Figure 8.15 Dependence of the flexural modulus (F) on the bulk density (ρ) of carbon–carbon composites carbonized at 1 000°C. From Ref. 5. (By permission of Pion, London.)

8.4 clearly shows that all composites (A, B, and C) are superior to the graphite (G) in Young's modulus, bending strength, tensile strength, fracture toughness, thermal shock resistance, and thermal shock fracture toughness. In particular, composite B is superior to composite A in all these properties. The load–elongation curves of composites A and B during tensile testing at various temperatures up to 2 400°C are shown in Figure 8.16 [65]. Composite B is inferior to composite C in tensile strength but is superior to composite C in fracture toughness above 1 500°C. Moreover, composite C shows laminar fracture during thermal shock, due to its two-dimensional layer structure, but composites A and B do not, as their felt reinforcement provides them with some degree of three-dimensional strength [65]. Three-dimensional weaving [66] of the carbon fibers can also be used to enhance the three-dimensional strength.

Table 8.4 Mechanical properties of three carbon–carbon composites (A, B, C) and a graphite (G). From Ref. 65.

		A^a	B^b	C^c	G
Bulk density (g/cm³)		1.68	1.77	1.57	1.76
Young's modulus (GPa)		13.5	26.3	17.0	10.5
Vickers hardness[d] (MPa)		135	163	–	172
Bending strength (MPa)		65.7	96.9	–	39.6
Tensile strength (MPa)	R.T.	35.7	55.4	68	28
	800°C	43.4	65.4	88	30
	1 600°C	42.0	50.4	102	37
	2 400°C	62.7	83.0	111	44
Fracture toughness (MPa.m$^{1/2}$)	R.T.	2.96	3.44	4.0	0.8
	800°C	2.82	3.58	5.5	0.8
	1 600°C	4.64	6.75	6.1	1.0
	2 400°C	5.30	12.9	7.0	1.9
Thermal diffusivity (mm²/sec)		62.4	56.6	–	48.0
Thermal shock resistance (W/mm)		≈148	≈155	≈171	50 ± 6
Thermal shock fracture toughness (W/mm$^{1/2}$)		≈779	≈805	≈856	33 ± 3

[a]Pitch carbon–carbon composite.
[b]PAN carbon–carbon composite.
[c]Two-dimensional rayon carbon–carbon composite.
[d]Load 5 kg.

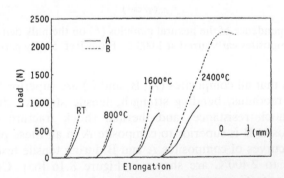

Figure 8.16 Load–elongation curves during tensile testing of carbon–carbon composites at various temperatures. From Ref. 65. (Reprinted with permission from Pergamon Press plc.)

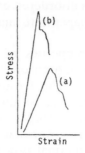

Figure 8.17 Tensile stress–strain curves of carbon–carbon composites made with pitch-based graphitized fabric: (a) carbonized composite, and (b) graphitized composite. From Ref. 67. (By permission of the publishers, Butterworth–Heinemann Ltd.)

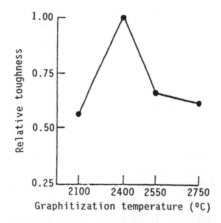

Figure 8.18 Relative toughness as a function of graphitization temperature of a three-dimensional carbon–carbon composite with a pitch-based carbon matrix. From Ref. 26. (Reprinted by permission of Kluwer Academic Publishers.)

Heat treatment temperature has a significant effect on the mechanical properties of carbon–carbon composites. Composites carbonized at 1 000°C upon subsequent graphitization at 2 700°C show a 54% increase in the flexural strength, a 40% decrease in the interlaminar shear strength, and a 93% increase in the flexural modulus [67]. The tensile stress–strain curves before and after graphitization are shown in Figure 8.17 [67]. This suggests that the fiber–matrix interaction is different before and after graphitization. The increase in the flexural modulus is probably due to the further graphitization of the carbon fibers under the influence of the carbon matrix around them, even though the fibers have been graphitized prior to this [67]. The choice of the graphitization temperature affects the toughness of the composite, as shown in Figure 8.18. For a pitch-based matrix, the optimum graphitization temperature is 2 400°C, where the microstructure is sufficiently ordered to accommodate some slip from shear forces but is disordered enough to prevent long-range slip [26]. A similar graphitization temperature may be used for a polymer-based matrix [68].

The tensile and flexural properties of carbon–carbon composites are fiber-dominated, whereas the compression behavior is mainly affected by density and matrix morphologies. The tensile moduli are sometimes higher than the values calculated according to the fiber content because of the contribution of the sheath matrix morphology (stress-graphitization of the matrix). Tensile strength levels are lower than the calculated values due to the residual stresses resulting from thermal processing. A high glass-like carbon fraction in the matrix is associated with enhanced strength and modulus, both in tension and compression [69].

The effect of surface treatment of the carbon fibers is significant on the mechanical properties of the resulting carbon–carbon composites. Consider the case of surface treatment using nitric acid. Figure 8.19 shows the effect of the treatment time on the flexural strength of polymer composites, carbonized composites (1 000°C heat treatment), and graphitized composites (2 700°C heat treatment) [67]. For both polymer and carbonized composites, the flexural strength initially increases with treatment time and drops with further treatment; the initial increase is smaller and the later drop is much more for carbonized composites than for polymer composites. This means that surface-treated fibers having strong bonding with the polymer exhibit high flexural strength in polymer composites, but result in carbonized composites of poor flexural strength. For graphitized composites, the flexural strength increases monotonically with increasing treatment time. Graphitization causes composites with surface-treated fibers to increase in flexural strength and interlaminar shear strength and those with untreated fibers to decrease in flexural strength and interlaminar shear strength. Hence, the fiber–matrix bonding is very poor in graphitized composites containing untreated fibers and is stronger in graphitized composites containing treated fibers [67].

The mechanical properties, density, and open porosity of carbon–carbon composites before and after graphitization are shown in Table 8.5 [12] for composites containing 50 vol.% fibers (PAN-based high modulus-type) and for matrices based on resin and CVI carbon. For the case of a resin-based matrix, the effect of surface treatment by wet oxidation of the fibers is also shown in Table 8.5. For the case of a CVI carbon matrix, the fibers were heated at above 1 300°C before CVI, so the functional groups on the fibers were eliminated. Consistent with Figure 8.19, Table 8.5 shows that the flexural strength is higher after graphitization for the case of surface-treated fibers, but is lower after graphitization for the case of nonsurface-treated fibers. Furthermore, Table 8.5 shows that the open porosity is not affected by graphitization for the case of surface-treated fibers, but is increased by graphitization for the case of nonsurface-treated fibers. In contrast, both flexural strength and open porosity are not affected by graphitization for the case of the CVI carbon matrix.

Table 8.5 Density, porosity, and mechanical properties of various types of carbon–carbon composites before and after graphitization. From Ref. 12.

| | Resin-based matrix | | CVI matrix |
	With surface-treated fiber	With nonsurface-treated fiber	
1. Carbonized			
Bulk density (g/cm³)	1.55	–	1.79
Open porosity (%)	4.5	7.0	13.0
Flexural strength (MPa)	150	900	520
2. Graphitized			
Open porosity (%)	4.5	10.5	12.0
Flexural strength (MPa)	600	350	500

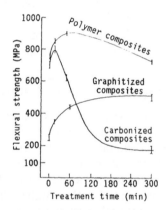

Figure 8.19 Variation of the flexural strength of composites with the surface treatment time of graphitized fibers used to make the composites. From Ref. 67. (By permission of the publishers, Butterworth–Heinemann Ltd.)

After carbonization, a residual stress remains at the interface between the fibers and the matrix. This is because, during carbonization, fiber–matrix interaction causes the crystallite basal planes to be aligned parallel to the fiber axis [61]; the resulting Poisson's effect elongates the fibers along the fiber axis and compresses them in the radial direction. This effect is indicated during carbonization by the transverse shrinkage and the longitudinal expansion of the composite. Its transverse shrinkage is larger than the shrinkage of the resin alone (which is the matrix material), while its longitudinal expansion is larger than the expansion of the fibers alone [12]. The residual stress can cause warpage [70].

The high-temperature resistance of carbon–carbon composites containing boron or zirconium diboride glass-forming oxidation inhibitors can be impaired by the reactions between the inhibitors and the carbonaceous components of the composite. These reactions, which probably form carbides, affect both fibers and matrix. They result in almost complete crystallization of the composite components. This crystallization transforms the microstructure of the composite, weakening it and producing brittle failure behavior. For boron, the reaction occurs at temperatures between 2 320 and 2 330°C; for zirconium diboride, it occurs at temperatures between 2 330 and 2 350°C [71].

For two-dimensional carbon–carbon composites containing plain weave fabric reinforcements under tension, the mode of failure of the fiber bundles depends on their curvature. Fiber bundles with small curvatures fail due to tensile stress or due to a combination of tensile and bending stresses. Fiber bundles with large curvatures fail due to shear stresses at the point where the local fiber direction is most inclined to the applied load [72].

The carbon matrix significantly enhances the carbon fibers' resistance to creep deformation due to the ability of the matrix to distribute loads more evenly and to impose a plastic flow-inhibiting, triaxial stress state in the fibers [73]. The thermally activated process for creep is controlled by vacancy formation and motion. For steady-state creep, the apparent activation energy is 1 082 kJ/mol, while the stress exponent has a value of 7.5 [74].

Thermal Conductivity and Electrical Resistivity

Carbon–carbon composites with high thermal conductivity are important for first wall components for nuclear fusion reactors, hypersonic aircraft, missiles and spacecraft, thermal radiator panels, and electronic heat sinks.

The thermal conductivity of carbon–carbon composites at < 1 000°C increases with the heat treatment temperature, particularly above 2 800°C [75], as more graphitic carbon is associated with a higher thermal conductivity. The thermal conductivity and electrical resistivity of two-dimensional weave carbon–carbon composites parallel and perpendicular to the laminates are shown in Table 8.6 [26]. Parallel to the fiber axis, the thermal conductivity is high and the electrical resistivity is low. Perpendicular to the laminate, the thermal conductivity is low and the electrical resistivity is high. Graphitization increases the thermal conductivity and decreases the electrical resistivity. The effect of heat treatment at 3 000°C for 1 h. on the thermal conductivity of a one-dimensional carbon–carbon composite is shown in Figure 8.20 [76], which also shows that the thermal conductivity decreases significantly with increasing test temperature. Of most interest is the fact that, after the 3 000°C heat treatment, the thermal conductivity is 500 W/m/K at 300°C [76], compared to a value of 225 W/m/K for Al and a value of 363 W/m/K for Cu at the same temperature. The above result on carbon–carbon composites is for ones made from Mitsubishi Kasei (Tokyo, Japan) K139 carbon fibers, which have a thermal conductivity of 107 W/m/K at room temperature. In contrast, vapor grown carbon fibers have a thermal conductivity of 1 900 W/m/K at 25°C [76]. Hence, carbon–carbon composites using vapor grown carbon fibers may have a

Table 8.6 Thermal conductivity and electrical resistivity parallel and perpendicular to the laminates of two-dimensional weave carbon–carbon composites. From Ref. 26.

Heat treatment temperature (°C)	Thermal conductivity (W/m/K) ∥	⊥	Electrical resistivity (μΩ.m) ∥	⊥
1 200	36–43	4–7	33–37	98–114
2 800	127–134	39–46	8–12	68–81

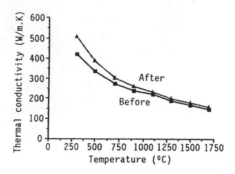

Figure 8.20 Thermal conductivity versus test temperature for one-dimensional carbon–carbon composites before and after heat treatment at 3 000°C. From Ref. 76.

thermal conductivity exceeding 1 000 W/m/K [77]. The low density of carbon makes the specific thermal conductivity of carbon–carbon composites outstandingly high compared to other materials. The use of porous carbon–carbon composites with even lower densities [78] may further increase the specific thermal conductivity.

As is the case for graphite, carbon–carbon composites are very low in thermal conductivity at temperatures less than 10 K. On the other hand, carbon–carbon composites are mechanically much stronger than graphite. Therefore, they are useful for rigid optical assemblies at low temperatures [79].

Vibration Damping Ability

Vibration damping is important to aerospace structures. Compared to the precursor polymer-matrix composite, a carbon–carbon composite has a lower resonance frequency and a higher damping ratio [80]. This is attributed to the transverse cracks, debonding, and porosity development of the matrix precursor during carbonization [80]. The fiber–matrix interface of a carbon–carbon composite is microfissured; numerous microcracks exist both within the matrix and along partially bonded interfaces. The microcracks within the matrix are

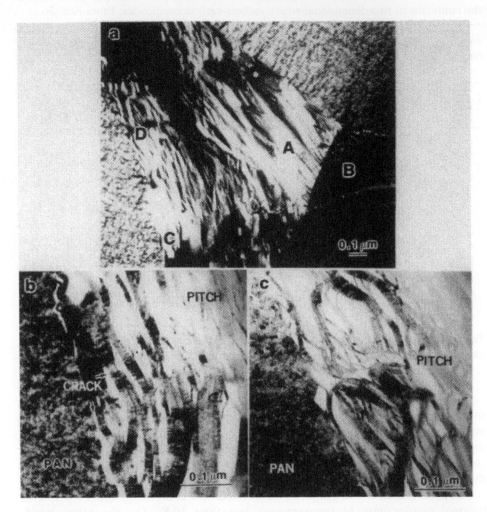

Figure 8.21 Transmission electron micrographs of a carbon–carbon composite made from PAN-based fibers and a mesophase pitch matrix, showing the fiber–matrix interface. (a) Dark field image showing mesophase domains A and B with different crystallite orientations in the intrabundle matrix region, such that the graphitic platelets in each domain are aligned roughly parallel to the nearest fiber surface. (b) Higher magnification bright field image of the region marked "C" in (a), revealing microcracks along and near the fiber–matrix interface, and between graphitic crystallite platelets. (c) Bright field image of the region marked "D" in (a), showing matrix crystallites that are thinner, more random and more bent than those in (b), probably due to the flow nature during mesophase transition. From Ref. 81. (By permission of Chapman & Hall, London.)

formed between and parallel to the basal planes of the graphite platelets such that they get smaller and denser near the fiber–matrix interface, as shown by transmission electron microscopy in Figure 8.21 [81].

Applications

Applications for carbon–carbon composites include aircraft brakes [82,83], heat pipes [84], reentry vehicles [85], rocket motor nozzles [85], hip replacements [86], biomedical implants [87–89], tools and dies [89], engine pistons [89], tiles for plasma facing armor [90], and electronic heat sinks [91].

References

1. R. Fujiura, T. Kojima, K. Kanno, I. Mochida, and Y. Korai, *Carbon* **31**(1), 97–102 (1993).
2. R.B. Sandor, in *Proc. Int. SAMPE Tech. Conf., 22, Advanced Materials: Looking Ahead to the 21st Century*, edited by L.D. Michelove, R.P. Caruso, P. Adams, and W.H. Fossey, Jr., 1990, pp. 647–657; *SAMPE Q.* **22**(3), 23–28 (1991).
3. J. Economy, H. Jung, and T. Gogeva, *Carbon* **30**(1), 81–85 (1992).
4. G. Savage, *Met. Mater. (Inst. Met.)* **4**(9), 544–548 (1988).
5. B. Rhee, S. Ryu, E. Fitzer, and W. Fritz, *High Temp.–High Pressures* **19**(6), 677–686 (1987).
6. J. Chlopek and S. Blzewicz, *Carbon* **29**(2), 127–131 (1991).
7. O.P. Bahl, L.M. Manocha, G. Bhatia, T.L. Dhami, and R.K. Aggarwal, *J. Sci. Ind. Res.* **50**(7), 533–538 (1991).
8. H.A. Aglan, *Int. SAMPE Symp. Exhib.*, **36**(2), 2237–2248 (1991).
9. A.J. Hosty, B. Rand, and F.R. Jones, *Inst. Phys. Conf. Ser., Vol. 111 New Materials and Their Applications 1990*, IOP, Bristol, U.K. and Philadelphia, 1990, pp. 521–530.
10. G. Gray and G.M. Savage, *Materials at High Temperatures* **9**(2), 102–109 (1991).
11. T. Hosomura and H. Okamoto, *Mater. Sci. Eng.* **A143**(1–2), 223–229 (1991).
12. S. Kimura, Y. Tanabe, and E. Yasuada, in *Proc. 4th Japan–U.S. Conf. Compos. Mater., 1988*, Technomic, Lancaster, PA, 1989, pp. 867–874.
13. E. Yasuda, Y. Tanabe, and K. Taniguchi, *Rep. Res. Lab. Eng. Mater., Tokyo Inst. Technol.* **13**, 113–119 (1988).
14. S. Marinkovic and S. Dimitrijevic, *Carbon* **23**(6), 691–699 (1985).
15. J.W. Davidson, in *Proc. Metal and Ceramic Matrix Composite Processing Conf., Vol. II*, U.S. Dept. of Defense Information Analysis Centers, 1984, pp. 181–185.
16. V. Markovic, *Fuel* **66**(11), 1512–1515 (1987).
17. P.M. Sheaffer and J.L. White, U.S. Patent 4,986,943 (1991).
18. A.J. Hosty, B. Rand, and F.R. Jones, *Inst. Phys. Conf. Ser., Vol. 111, New Materials and Their Applications 1990*, IOP, Bristol, U.K. and Philadelphia, 1990, pp. 521–530.
19. I. Charit, H. Harel, S. Fischer, and G. Marom, *Thermochim. Acta* **62**, 237–248 (1983).
20. T. Chang and A. Okura, *Trans. Iron Steel Inst. Jpn.*, **27**(3), 229–237 (1987).

21. L.M. Manocha and O.P. Bahl, *Carbon* **26**(1), 13–21 (1988).
22. L.M. Manocha, O.P. Bahl, and Y.K. Singh, *Tanso* **140**, 255–260 (1989).
23. L.M. Manocha, O.P. Bahl, and Y.K. Singh, *Carbon* **29**(3), 351–360 (1991).
24. L.M. Manocha, *Composites (Guildford, U.K.)* **19**(4), 311–319 (1988).
25. W. Kowbel and C.H. Shan, *Carbon* **28**(2–3), 287–299 (1990).
26. W. Huettner, in *Carbon Fibers Filaments and Composites*, edited by J.L Figueiredo, C.A. Bernardo, R.T.K. Baker, and K.J. Huttinger, Kluwer Academic, Dordrecht, 1990, pp. 275–300.
27. P. Ehrburger and J. Lahaye, *High Temp.–High Pressures* **22**(3), 309–316 (1990).
28. P.K. Jain, O.P. Bahl, and L.M. Manocha, *SAMPE Q.* **23**(3), 43–47 (1992).
29. L.M. Manocha, O.P. Bahl, and Y.K. Singh, *Carbon* **27**(3), 381–387 (1989).
30. S. Takano, T. Kinjo, T. Uruno, T. Tlomak, and C.P. Ju, *Ceram. Eng. Sci. Proc.* **12**, 1914–1930 (1991).
31. R.J. Zaldivar and G.S. Rellick, *Carbon* **29**(8), 1155–1163 (1991).
32. R.J. Zaldivar, G.S. Rellick, and J.M. Yang, *J. Mater. Res.* **8**(3), 501–511 (1993).
33. G.S. Rellick, D.J. Chang, and R.J. Zaldivar, *J. Mater. Res.* **7**(10), 2798–2809 (1992).
34. R.J. Zaldivar, R.W. Kobayashi, and G.S. Rellick, *Carbon* **29**(8), 1145–1153 (1991).
35. R.J. Zaldivar, G.S. Rellick, and J.–M. Yang, *SAMPE J.* **27**(5), 29–36 (1991).
36. R.E. Yeager and S.C. Shaw, in *Proc. Metal and Ceramic Matrix Composite Processing Conf., Vol. II*, U.S. Dept. of Defense Information Analysis Centers, 1984, pp. 145–180.
37. K.S. Goto, K.H. Han, and G.R. St. Pierre, *Trans. Iron Steel Inst. Jpn.* **26**(7), 597–603 (1986).
38. P. Crocker and B. McEnaney, *Carbon* **29**(7), 881–885 (1991).
39. J.E. Sheehan, *Carbon* **27**(5), 709–715 (1989).
40. H.V. Johnson, U.S. Patent 1,948,382.
41. T.-M. Wu, W.-C. Wei, and S. Hsu, *Carbon* **29**(8), 1257–1265 (1991).
42. R.C. Dickinson, *Mater. Res. Soc. Symp. Proc.*, **125** (Materials Stability and Environmental Degradation), edited by A. Barkatt, E.D. Verink, Jr. and L.R. Smith, 1988, pp. 3–11.
43. F.J. Buchanan and J.A. Little, *Surf. Coat. Technol.* **46**(2), 217–226 (1991).
44. J.R. Strife and J.E. Sheehan, *Am. Ceram. Soc. Bull.* **67**(2), 369–374 (1988).
45. D.W. McKee, *Carbon* **25**(4), 551–557 (1987).
46. D.W. McKee, *Carbon* **24**(6), 737–741 (1986).
47. P.E. Gray, U.S. Patent 4,894,286 (1990).
48. T.D. Nixon and J.D. Cawley, *J. Am. Ceram. Soc.* **75**(3), 703–708 (1992).
49. P.E. Gray, U.S. Patent 4,937,101 (1990).
50. L.M. Niebylski, U.S. Patent 4,910,173 (1990).
51. D.W. McKee, *Carbon* **26**(5), 659–665 (1988).
52. P. Ehrburger, in *Carbon Fibers Filaments and Composites*, edited by J.L. Figueiredo, C.A. Bernardo, R.T.K. Baker, and K.J. Huttinger, Kluwer Academic, Dordrecht, 1990, pp. 327–336.
53. P. Ehrburger, P. Baranne, and J. Lahaye, *Carbon* **24**(4), 495–499 (1986).
54. B. Bavarian, V. Arrieta, and M. Zamanzadeh, in *Proc. Int. SAMPE Symp. and Exhib., 35, Advanced Materials: Challenge Next Decade*, edited by G. Janicki, V. Bailey, and H. Schjelderup, 1990, pp. 1348–1362.
55. T.-M. Wu, W.-C. Wei, and S. Hsu, *Carbon* **29**(8), 1257–1265 (1991).
56. K.H. Han, H. Ono, K.S. Goto, and G.R. St. Pierre, *J. Electrochem. Soc.* **134**(4), 1003–1009 (1987).

57. E.J. Hippo, N. Murdie, and W. Kowbel, *Carbon* **27**(3), 331–336 (1989).
58. C.T. Ho and D.D.L. Chung, *Carbon* **28**(6), 815–824 (1990).
59. E.J. Hippo, N. Murdie, and A. Hyjazie, *Carbon* **27**(5), 689–695 (1989).
60. R.A. Meyer and S.R. Gyetvay, *ACS Symp, Ser., Vol 303, Petroleum-Derived Carbons*, American Chemical Society, Washington, D.C., 1986, pp. 380–394.
61. L.H. Peebles, Jr., R.A. Meyer, and J. Jortner, in *Proc. 2nd Int. Conf. Compos. Interfaces, Interfaces Polym., Ceram., Met. Matrix Compos.*, edited by H. Ishida, Elsevier, New York, 1988, pp. 1–16.
62. L.E. Jones, P.A. Thrower, and P.L. Walker, Jr., *Carbon* **24**(1), 51–59 (1986).
63. J.X. Zhao, R.C. Bradt, and P.L. Walker, Jr., *Carbon* **23**(1), 9–13 (1985).
64. H. Aglan, *J. Mater. Sci. Lett.* **11**(4), 241–243 (1992).
65. S. Sato, A. Kurumada, H. Iwaki, and Y. Komatsu, *Carbon* **27**(6), 791–801 (1989).
66. H. Weisshaus, S. Kenig, E. Kastner, and A. Siegmann, *Carbon* **28**(1), 125–135 (1990).
67. L.M. Manocha, O.P. Bahl, and Y.K. Singh, in *Proc. Int. Conf. Interfacial Phenomena in Composite Materials '89*, edited by F.R. Jones, Butterworth, 1989, pp. 310–315.
68. R.B. Sandor, *SAMPE Q.* **22**(3), 23–28 (1991).
69. H. Weisshaus, S. Kenig, and A. Siegmann, *Carbon* **29**(8), 1203–1220 (1991).
70. L.A. Feldman, *J. Mater. Sci. Lett.* **5**, 1266–1268 (1986).
71. S. Ragan and G.T. Emmerson, *Carbon* **30**(3), 339–344 (1992).
72. P.B. Pollock, *Carbon* **28**(5), 717–732 (1990).
73. G. Sines, Z. Yang, and B.D. Vickers, *J. Am. Ceram. Soc.* **72**(1), 54–59 (1989).
74. G. Sines, Z. Yang, and B.D. Vickers, *Carbon* **27**(3), 403–415 (1989).
75. R.B. Dinwiddie, T.D. Burchell, and C.F. Baker, *Ext. Abstr. Program–Bienn. Conf. Carbon* **20**, 642–643 (1991).
76. J.W. Sapp, Jr., D.A. Bowers, R.B. Dinwiddie, and T.D. Burchell, *Ext. Abstr. Program – Bienn Conf. Carbon* **20**, 644–645 (1991).
77. M.L. Lake, J.K. Hickok, K.K. Brito, and L.L. Begg, in *Proc. Int. SAMPE Symp. Exhib.*, **35**, *Advanced Materials: Challenge Next Decade*, 960–969, 1990.
78. X. Shui and D.D.L. Chung, *Ext. Abstr. Program—Bienn. Conf. Carbon* **20**, 376–377 (1991).
79. C. Blondel, R. Roquessalane, O.A. Testard, F. Latimier, and D. Viratelle, *Cryogenics* **29**(5), 569–571 (1989).
80. U.K. Vaidya, P.K. Raju, and W. Kowbel, *Carbon* **30**(6), 925–929 (1992).
81. C.P. Ju, J. Don, and P. Tlomak, *J. Mater. Sci.* **26**(24), 6753–6758 (1991).
82. S. Awasthi and J.L. Wood, *Ceram. Eng. Sci. Proc.* **9**(7–8), 553–560 (1988).
83. S. Awasthi and J.L. Wood, *Adv. Ceram. Mater.* **3**(5), 449–451 (1988).
84. R.D. Rovang, R.B. Dirling, Jr., and R.A. Holzl, in *Proc. 25th Intersoc. Energy Convers. Eng. Conf., Vol. 2*, 1990, pp. 147–150.
85. Y. Grenie, in *Materials Science Monograph 41, Looking Ahead for Materials and Processes, Proc. 8th Int. Conf. Society for the Advancement of Material and Process Engineering, European Chapter, La Baule, Loire-Atlantique, France, 1987*, edited by J. De Bossu, G. Briens, and P. Lissac, Elsevier, Amsterdam, 1987. pp. 377–386.
86. J.A. Oliver, *R&D Magazine*, December 1991, pp. 17–18.
87. D. Adams and D.F. Williams, *Biomaterials (Guildford, U.K.)* **5**(2), 59–64 (1984).
88. N. More, C. Baquey, M.F. Harmand, R. Duphil, F. Rouais, M. Trinquecoste, and A. Marchand, *Adv. Biomater.*, **6**, 343–348 (1986).

89. E. Fitzer, *Carbon* **25**(2), 163–190 (1987).
90. K. Ioki, K. Namiki, S. Tsujimura, M. Toyoda, M. Seki, and H. Takatsu, *Fusion Eng. Des.* **15**(1), 31–38 (1991).
91. W.H. Pfeifer, J.A. Tallon, W.T. Shih, B.L. Tarasen, and G.B. Engle, in *Proc. 6th Int. SAMPE Electron. Conf.*, 1992, pp. 734–747.

CHAPTER 9

Ceramic-Matrix Composites

Ceramic-matrix carbon fiber composites are gaining increasing attention because (1) the good oxidation resistance of the ceramic matrix (compared to a carbon matrix) makes the composites attractive for high-temperature applications (e.g., aerospace and engine components), and (2) the continuously decreasing price of carbon fibers makes the use of carbon fiber reinforced concrete economically feasible.

The fibers serve mainly to increase the toughness and strength (tensile and flexural) of the composite due to their tendency to be partially pulled out during the deformation. This pullout absorbs energy, thereby toughening the composite. Although the fiber pullout has its advantage, the bonding between the fibers and the matrix must still be sufficiently good in order for the fibers to strengthen the composite effectively. Therefore, the control of the bonding between the fibers and the matrix is important for the development of these composites.

A second function of the carbon fibers is to decrease the electrical resistivity of the composite, as the ceramic is electrically insulating and the carbon fibers are electrically conducting [1,2]. Electrically conducting concrete is useful as a smart structure material that allows nondestructive flaw detection by electrical probing, as a crack will cause the electrical resistance to increase in its vicinity [3]. Related to a decrease in the electrical resistivity is an increase in the electromagnetic interference (EMI) shielding effectiveness [4], which is of growing importance due to the abundance and sensitivity of modern electronics.

A third function of the carbon fibers is to increase the thermal conductivity of the composite, as the ceramic is mostly thermally insulating whereas the carbon fibers are thermally conductive. In carbon fiber reinforced concrete the enhanced thermal conductivity is attractive for heat conduction in the overlay of heated bridges and in the interior of buildings heated by radiation heating. In electronic, aerospace, and engine components the enhanced thermal conductivity is attractive for heat dissipation.

A fourth function of the carbon fibers is to decrease the drying shrinkage in the case of ceramic matrices prepared by using slurries or slips [1]. In general, the drying shrinkage decreases with increasing solid content in the slurry. Carbon fibers are a particularly effective solid for decreasing the drying shrinkage, probably because of their high aspect ratio and stiffness. This function is attractive for the dimensional control of parts (e.g., pothole repair material) made from the composites.

Ceramic-matrix carbon fiber composites that have been reported fall into the following categories: (1) carbon fiber reinforced concretes and gypsums, (2) carbon fiber reinforced glasses, and (3) other carbon fiber reinforced ceramics (MgO, Al_2O_3, SiC, mullite). The choice of the ceramic matrix depends on the application and on the chemical compatibility of the matrix with the carbon fibers. As the oxidation of carbon is catalyzed by an alkaline environment, an alkaline matrix is not preferred. Nevertheless, carbon fibers are used in concrete, which is alkaline, because competitive fibers, such as glass fibers and polymer fibers, are even less stable in concrete. There has been no report of carbon fiber degradation in concrete. Matrices that react with carbon to form carbides are to be avoided, especially in high-temperature applications, as such reactions will degrade the carbon fibers.

Carbon Fiber Reinforced Concretes and Gypsums

Due to the low prices of concretes and gypsums and the continuously decreasing price of carbon fibers, carbon fibers are only recently gaining importance as a reinforcement for these materials. The carbon fibers used are usually short, isotropic, and pitch-based, as they are the least expensive among carbon fibers, even though their mechanical properties are poor compared to other carbon fibers. The use of carbon fibers in concrete is still in its infancy, mainly in Japan and Europe, but has not begun in the United States. Once carbon fibers become widely used for concrete, the price of the pitch-based carbon fibers will greatly decrease, because of the lower production cost per pound of the fibers when the production volume is large. (Pitch itself is cheap.) Therefore, the economics of using carbon fibers in concrete will become even more attractive. The use of carbon fibers in gypsum has not yet begun industrially.

Carbon fibers are attractive for concretes and gypsums because of their superior chemical stability compared to glass fibers (which tend to dissolve away in the alkaline environment of concrete or gypsum) and polymer fibers (e.g. polypropylene fibers, acrylic fibers). Even though the oxidation of carbon is catalyzed by an alkaline environment, the chemical stability of carbon in concrete is apparently sufficient, as indicated by the carbon fiber reinforced concrete monument built in Japan in 1982. Another type of fiber is asbestos, but asbestos fibers are no longer desirable because of their medical hazard. Yet another type of fiber is steel fibers, but steel fibers used for concrete are large; they are more like nails than fibers. Steel fibers that are really fine, like typical fibers, are very expensive compared to carbon fibers.

The effect of carbon fibers on the properties of concrete (with stones) or mortar (without stones) include the following:

- increased flexural strength
- increased flexural toughness (area under the curve of flexural stress versus displacement)
- increased durability under cyclic loading
- decreased compressive strength
- increased air content
- improved freeze–thaw durability
- decreased drying shrinkage
- decreased electrical resistivity
- increased electromagnetic interference (EMI) shielding effectiveness
- increased thermal conductivity
- improved resistance to earthquake damage

The only negative effect is the decrease in the compressive strength, but this decrease can be more than compensated by the addition of chemical agents and silica fume. The increased flexural strength and flexural toughness are most attractive, because concrete is inherently weak under flexure but strong under compression. Related to the increased flexural strength and flexural toughness is the improved resistance to earthquake damage. The improved freeze–thaw durability is related to the increased air content.

The magnitudes of the above effects increase with increasing fiber content. However, a high fiber content is economically not attractive; it makes the fiber dispersion (concrete processing) more difficult and the concrete mix less workable. This difficulty causes the flexural strength to decrease when the fiber content exceeds an optimum value, while the flexural toughness keeps increasing with increasing fiber content. The minimum fiber content for the increase in flexural strength to be appreciable is 0.1 vol.% [1]. The optimum fiber content for the highest flexural strength depends on the dispersant used.

The dispersion of the carbon fibers is important for their effectiveness. This can be achieved in the dry state, known as dry mixing, or in the wet state, known as wet mixing. Dry mixing involves mixing the fibers with the fine aggregate by hand (i.e., by roughly adding alternately a fiber layer and a fine aggregate layer to the mix), subsequently adding cement and silica fume, then mixing by using a Hobart mixer for ~ 5 min. Chemical agents such as a water-reducing agent and accelerating agents may be further added and mixed with the Hobart mixer. After that, the mixture is poured into a stone concrete mixer, to which the coarse aggregate is added. The dry mixing procedure is tedious because of the first step, which is by hand. A much more practical mixing procedure is wet mixing, which involves using a dispersant dissolved in water and adding the fibers to the solution. The dispersant can be methylcellulose (in the amount of 0.4% of the weight of the cement, provided by Dow Chemical, Midland, MI, as Methocel A15-LV), which unfortunately introduces foam, so that a defoamer, such as Colloids 1010 (Colloids, Marietta, GA, in

Table 9.1 Flexural strength of mortars. The top group is without sand; the bottom group is with sand. From Ref. 5.

	Flexural strength (MPa)		
Samples	With L	With M	With M + SF
Plain	2.24($\pm$3.2%)	2.24($\pm$3.2%)	1.53($\pm$2.1%)
+ L/ + M/ + M + SF	3.62($\pm$4.2%)	2.29($\pm$3.2%)	2.79($\pm$2.2%)
+ 0.53 vol.% F	7.62($\pm$5.6%)	3.68($\pm$4.1%)	3.60($\pm$3.6%)
+ 1.06 vol.% F	7.72($\pm$7.8%)	5.65($\pm$7.8%)	5.60($\pm$4.9%)
+ 2.12 vol.% F	9.03($\pm$2.8%)	8.58($\pm$4.8%)	9.51($\pm$3.6%)
+ 3.18 vol.% F	7.42($\pm$7.1%)	9.21($\pm$5.4%)	11.82($\pm$4.4%)
+ 4.24 vol.% F	7.62($\pm$4.3%)	9.60($\pm$6.9%)	12.22($\pm$6.3%)
Plain + sand	4.36($\pm$2.2%)	3.70($\pm$1.9%)	3.70($\pm$1.9%)
+ L/ + M/ + M + SF	6.02($\pm$4.8%)	3.74($\pm$2.4%)	3.94($\pm$3.4%)
+ 0.35 vol. % F	8.24($\pm$9.2%)	4.95($\pm$4.2%)	5.27($\pm$4.6%)
+ 0.70 vol. % F	8.84($\pm$7.2%)	5.29($\pm$2.9%)	5.63($\pm$4.4%)
+ 1.40 vol. % F	9.14($\pm$4.5%)	6.14($\pm$4.6%)	7.69($\pm$3.2%)
+ 2.10 vol. % F	9.26($\pm$7.7%)	7.28($\pm$5.2%)	9.87($\pm$4.9%)
+ 2.80 vol. % F	8.07($\pm$5.5%)	6.66($\pm$7.2%)	9.52($\pm$6.3%)
+ 3.50 vol. % F	7.92($\pm$5.1%)	6.60($\pm$4.9%)	9.51($\pm$5.9%)
+ 4.20 vol. % F	8.10($\pm$9.2%)	5.81($\pm$7.8%)	9.40($\pm$8.2%)

L = latex; M = methylcellulose; SF = silica fume; F = fibers.

the amount of 0.13 vol.% of the concrete) must also be used. After this, the fine aggregate, cement and silica fume are added and the mixture is stirred with a Hobart mixer. The rest of the procedure is similar to dry mixing. Dry mixing and wet mixing give carbon fiber reinforced concretes of similar flexural strength. Instead of methylcellulose, latex may be used as the dispersant, but the latex amount required is large (20% of the weight of the cement), thus making the formulation too expensive.

Silica fume helps the dispersion of the fibers. In addition, it reduces the porosity of the concrete, thereby giving the fibers a better chance of bonding to the concrete. The preferred kind of silica fume is that with a high SiO_2 content (e.g., 94%) and a high surface area (e.g., $22 \, m^2/g$). It may be used in the amount of 0.15 of the weight of the cement.

Tables 9.1 and 9.2 [5] list the flexural strength and flexural toughness of mortars without sand (no aggregate) and mortars with sand for (1) latex, (2) methylcellulose, and (3) methylcellulose plus silica fume as the dispersants. Figures 9.1 and 9.2 show the curves of flexural stress versus displacement for mortars without sand and with sand, respectively, for Case 2.

For achieving the highest flexural strength, the optimum carbon fiber volume fraction is 2.1% for mortar using latex as the dispersant, whether sand is present or not. The optimum carbon fiber volume fraction is $\geq$4.2% when

Table 9.2 Flexural toughness of mortars. The top group is without sand; the bottom group is with sand. From Ref. 5.

Samples	Flexural toughness (MPa.mm)		
	With L	*With M*	*With M + SF*
Plain	0.056	0.056	0.038
+ L/ + M/ + M + SF	0.202	0.105	0.193
+ 0.53 vol.% F	0.967	0.630	1.108
+ 1.06 vol.% F	1.197	1.360	1.584
+ 2.12 vol.% F	2.739	3.067	2.718
+ 3.18 vol.% F	4.151	4.384	6.108
+ 4.24 vol.% F	4.627	5.574	12.305
Plain + sand	0.223	0.189	0.189
+ L/ + M/ + M + SF	0.500	0.191	0.200
+ 0.35 vol.% F	0.856	0.614	1.374
+ 0.70 vol.% F	1.170	1.057	1.484
+ 1.40 vol.% F	1.849	2.398	2.274
+ 2.10 vol.% F	2.186	2.555	3.808
+ 2.80 vol.% F	3.135	2.915	3.941
+ 3.50 vol.% F	3.716	2.989	4.545
+ 4.20 vol.% F	5.013	3.396	5.577

L = latex; M = methylcellulose; SF = silica fume; F = fibers.

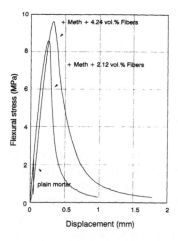

Figure 9.1 Flexural stress versus displacement during three-point bending (ASTM C348-80, span = 140 mm, specimen size = 40 × 40 × 160 mm) at 7 days of curing of mortars without sand.

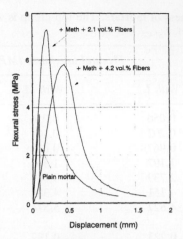

Figure 9.2 Flexural stress versus displacement during three-point bending (ASTM C348-80, span = 140 mm, specimen size = 40 × 40 × 160 mm) at 7 days of curing of mortars with sand.

either methylcellulose or methylcellulose plus silica fume is used as the dispersant for mortar without sand, and is 2.1% for mortar with sand. For achieving the highest flexural toughness, the optimum carbon fiber volume fraction is ≥ 4.2%, no matter what kind of dispersant is used and whether sand is present or not. When the fiber content is less than 2 vol.%, the highest flexural strength can be obtained by using latex as the dispersant, whether with or without sand. For fiber contents higher than 2 vol.%, the highest flexural strength is obtained by using both methylcellulose and silica fume as the dispersant.

Mortars containing latex as the dispersant are the most expensive among the various types of mortars in Tables 9.1 and 9.2 because of the cost of the large amount of latex used. If smaller amounts of latex were used, the mortars would not compete in performance with those using the other two dispersants. Since the fractional increases in flexural strength and toughness of the mortar containing methylcellulose plus silica fume are always higher than those of the mortar containing methylcellulose, it can be concluded that methylcellulose plus silica fume as the dispersant is most attractive among the three dispersants studied.

For the case without sand, the highest flexural strength and toughness (12.2 MPa and 12.3 MPa.mm) are attained by the mortar containing methylcellulose plus silica fume and 4.2 vol.% carbon fibers; the fractional increases in the flexural strength and toughness are 7 000% and 32 000%, respectively, relative to plain mortar. For the case with sand, the highest flexural strength and toughness (9.9 MPa and 5.6 MPa.mm, respectively) are attained by the mortar containing methylcellulose plus silica fume and 2.1 vol.% and 4.2 vol.% carbon fibers, respectively; the fractional increase in the flexural strength and toughness are 170% and 2 900%, respectively, relative to plain mortar.

For fiber contents above 3 vol.%, the presence of sand decreases the workability and thus decreases the flexural strength and toughness when methylcellulose plus silica fume is the dispersant; for fiber contents below 3 vol.%, the presence of sand helps to increase the flexural strength and toughness when methylcellulose plus silica fume is the dispersant.

Carbon fibers have been used in mortars by a number of workers [7–11] other than this author, though the fractional increase in the flexural strength due to the fibers is low compared to the data in Table 9.1 for similar values of the fiber volume fraction and sand/cement ratio. The difference is attributed to differences in the techniques for mixing and fiber dispersion.

The magnitudes of the above effects also depend on the fiber length. In particular, the flexural strength increases with increasing fiber length, though long fibers are hard to disperse. The optimum fiber length is 5 mm prior to concrete mixing for the case of mixing using a Hobart mixer. During concrete mixing, the fiber length decreases by 40% in the case of using a Hobart mixer (like a cake mixer) with a flat beater for mixing everything minus the coarse aggregate, followed by using a stone concrete mixer for mixing everything. The length decrease occurs in the Hobart mixer, not the stone concrete mixer. Specially designed mixers are available in Japan for more gentle mixing than provided by the Hobart mixers. However, the specially designed mixers require a long mixing time and add considerable cost to the concrete processing. By the use of dispersants for the carbon fibers, no special mixer (e.g., Hobart) is required.

The effectiveness of the carbon fibers, of length 5 mm before the mixing (with the Hobart mixer), in increasing the flexural strength does not vary much with the aggregate size ranging from ≤2 mm to 25 mm [1]. Thus, carbon fibers can be used in mortars (with fine aggregate only) as well as concretes (with both fine and coarse aggregates).

The water-reducing agent increases the fluidity of the concrete mix. Its importance increases with the carbon fiber content, as the carbon fibers decrease the fluidity of the concrete mix. The water-reducing agent can be a sodium salt of a condensed naphthalenesulfonic acid (e.g., TAMOL SN of Rohm and Haas, Philadelphia, PA; 93–96% acid, 4–7% water) in the amount of 2% of the weight of the cement for the case of carbon fibers in the amount of 0.5% of the weight of the cement [1].

The accelerating agents reduce the porosity of the concrete, thereby making the fibers have a better chance of bonding to the concrete. They can be triethanolamine in the amount of 0.06% of the weight of the cement, potassium aluminum sulfate in the amount of 0.5% of the weight of the cement, together with sodium sulfate in the amount of 0.5% of the weight of the cement [1,6].

As an increase in the water/cement ratio in general lowers the strength of concretes, this ratio should be kept to a minimum while maintaining sufficient fluidity in the concrete mix. For concrete containing carbon fibers in the amount of 0.5% of the weight of the cement, a water/cement ratio of 0.50 is

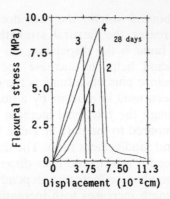

Figure 9.3 Flexural stress versus displacement during three-point bending (ASTM C348-80, span = 9 in., specimen size = 3 × 3 × 11 in.) at 28 days of curing of (1) plain concrete, (2) concrete containing carbon fibers (0.5% of the cement weight), (3) concrete containing silica fume, water-reducing agent, and accelerating agents, (4) concrete containing carbon fibers (0.5% of the cement weight), silica fume, water-reducing agent, and accelerating agents. Note that (4) has the largest area under the curve, so it has the highest flexural toughness. From Ref. 1.

needed in order for the fluidity to be sufficient [1]. Both the carbon fibers and the accelerating agent, sodium sulfate, decrease the fluidity of the concrete mix [1].

With economy in mind, carbon fibers in the amount of 0.5% of the weight of the cement are suitable in concrete. With a water/cement ratio of 0.5 and a cement:fine aggregate:coarse aggregate ratio of 1:1.5:2.49 by weight (fine aggregate size ≤ 5 mm; coarse aggregate size ≤ 25 mm), the fiber volume fraction is 0.19%, the flexural strength is increased by 85% and the flexural toughness (area under the curve of flexural stress versus deflection, Figure 9.3) is increased by 205% at 28 days of curing, compared to the corresponding values of plain concrete (without silica fume, water-reducing agent, or accelerating agents). Compared instead to concrete containing silica fume, water-reducing agent, and accelerating agents (no fiber), the flexural strength increase is 17% and the flexural toughness increase is 35%. On the other hand, the compressive strength is increased by 22% at 90 days of curing, compared to the corresponding value of plain concrete. Compared instead to concrete containing silica fume, water-reducing agent, and accelerating agents (no fiber), the compressive strength is decreased by 35%. Hence, carbon fibers decrease the compressive strength of concrete, though the decrease can be more than compensated by the use of silica fume, water-reducing agent, and accelerating agents. However, carbon fibers improve the ductility of the concrete in the lateral direction under axial compression (Figure 9.4) [1].

The origin of the decrease in the compressive strength due to the carbon fibers is that the fibers increase the air content, which is 1% for plain concrete, 7% for concrete with fibers only, 3% for concrete with silica fume and the

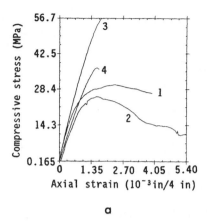
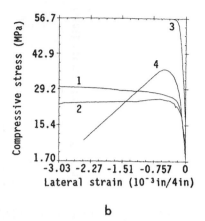

a   b

Figure 9.4   Compressive stress versus (a) axial strain and (b) lateral strain during axial compressive testing (ASTM C39-83b, specimen size = 4 in. diameter × 8 in. length) at 90 days of curing of (1) plain concrete, (2) concrete containing carbon fibers (0.5% of the cement weight), (3) concrete containing silica fume, water-reducing agent, and accelerating agents, (4) concrete containing carbon fibers (0.5% of the cement weight), silica fume, water-reducing agent, and accelerating agents. Note that (3) has the highest compressive strength. Comparison of (3) and (4) in Figure 9.4b shows that (4) is more ductile than (3) in the lateral direction; the lateral direction experiences tension during axial compression. From Ref. 1.

chemical agents, and 6% for concrete with fibers as well as silica fume and the chemical agents [1]. The effect of the fibers on the air content is associated with the foam generated by the dispersant.

Because of the increase of the air content, carbon fibers improve the freeze–thaw durability of concrete, even in the absence of an air entrainer [1]. Nevertheless, if an air entrainer is further used, the water/cement ratio can be decreased to 0.45 and the air content is 9% when fibers, silica fume, and chemical agents are present; this compares with a value of 6% for plain concrete containing the air entrainer. The excessively high air content of 9% makes the carbon fibers capable of increasing the flexural strength by only 79% and the flexural toughness by only 53% at 28 days of curing, compared to the corresponding values of plain concrete containing the air entrainer. These percentage increases are lower than those for the case without an air entrainer. Hence, the use of an air entrainer in the typical amount is not desirable.

The drying shrinkage is greatly decreased by the carbon fibers, as shown in Figure 9.5 [1]. The decrease is 90% at 14 days of curing, compared to plain concrete. This effect is attractive for concrete used for pothole repair, precasts, and other applications where dimensional control is needed.

The carbon fibers in the amount of 0.2 vol.% and other additives increase the material price of the concrete by 20–39%, relative to the price of plain concrete.

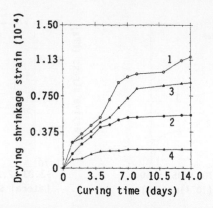

Figure 9.5 Drying shrinkage strain (ASTM C490-83a, specimen
size = $3 \times 3 \times 11.25$ in.) versus the curing time for (1) plain concrete, (2) concrete
containing carbon fibers (0.5% of the cement weight), (3) concrete containing silica
fume, water-reducing agent, and accelerating agents, (4) concrete containing carbon
fibers (0.5% of the cement weight), silica fume, water-reducing agent, and accelerating
agents. Note that (2) and (4) both contain carbon fibers and have less drying shrinkage
than (1) and (3), which contain no fibers. From Ref. 1.

Competitive fibers include organic and steel fibers. Acrylic fibers in the
amount of 2.5 vol.% increase the flexural strength by 28% and the flexural
toughness by 240% [12], whereas steel fibers in the amount of 1.2 vol.%
increase the flexural strength by 41% and the flexural toughness by 1 500%
[13]. In spite of the large fiber volume fractions, acrylic and steel fibers yield
fractional increases in the flexural strength that are lower than that of the
carbon fibers (0.2 vol.%). However, acrylic and steel fibers of such large
volume fractions yield fractional increases in the flexural toughness that are
higher than that of 0.2 vol.% carbon fibers. In addition to the better
effectiveness in increasing the flexural strength, carbon fibers are attractive for
their chemical stability.

Yet another attraction of carbon fibers lies in their low electrical
resistivity. For 3 mm long isotropic pitch-based carbon fibers in mortar with a
sand/cement ratio of 0.5 (by weight) and a water/cement ratio of 0.36 (by
weight), the electrical resistivity of the mortar is 8 730, 378, 93.7, 37.9, 22.3,
and 17.2 Ω.cm for carbon fibers in the amounts of 0.2, 0.5, 0.8, 1.0, 1.5, and
2.0% of the weight of the cement. respectively, at 23 days of curing [6]. The
resistivity increases slightly with increasing curing time, but the effect is small.
For example, for 3 mm fibers in the amount of 0.5% of the weight of the
cement, the resistivity is 355, 368, and 378 Ω.cm at 2, 5, and 23 days of curing
respectively [6]. The resistivity decreases with increasing fiber length, but the
decrease is appreciable only for fiber lengths ≤ 5 mm. For example, for fibers in
the amount of 0.5% of the weight of the cement, the resistivity is 113, 42.3,
46.8, and 14.2 Ω.cm for fiber lengths of 3.0, 5.1, 25.4, and 50.8 mm,
respectively, at 7 days of curing [6].

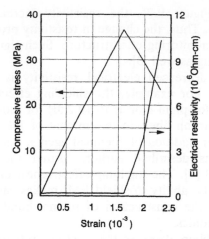

Figure 9.6 Variations of compressive stress and electrical resistivity with strain during static loading up to fracture for plain mortar without fibers.

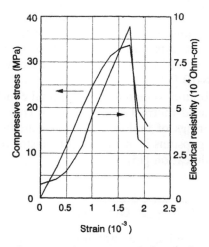

Figure 9.7 Variations of compressive stress and electrical resistivity with strain during static loading up to fracture for mortar with carbon fibers (0.24 vol.%) and methylcellulose.

The low electrical resistivity of carbon fiber reinforced concrete allows this concrete to function as a smart structure material which provides the ability for nondestructive in situ flaw detection during static or cyclic loading. Figures 9.6 and 9.7 show the variations of the compressive stress and of the electrical resistivity (along the stress direction) with the compressive strain during static loading up to fracture for plain mortar (without fibers) and for mortar containing 0.24 vol.% carbon fibers and methylcellulose, respectively. Without fibers (Figure 9.6), the resistivity did not vary during the deformation.

With carbon fibers (Figure 9.7), the resistivity increased as deformation occurred up to fracture. Thus, the change in resistivity provides a signature for nondestructive health monitoring of a structure. Such signatures occur reversibly during cyclic loading as well [3].

Along with the decreased electrical resistivity is the increased EMI shielding effectiveness. For example, the shielding effectiveness at 1 GHz is increased from 0.4 dB for plain mortar to 10.2 dB for mortar containing 3 mm isotropic pitch-based carbon fibers in the amount of 1% of the weight of the cement [4]. The further addition of chemical agents, namely the accelerating agents in the amounts described earlier, further increases the shielding effectiveness to 14.8 dB [4]. The shielding effectiveness increases with increasing fiber content [4]. The ability to shield EMI makes carbon fiber reinforced concrete attractive for the construction of rooms, buildings or underground vaults that house electronics.

Similarly carbon fibers decrease the electrical resistivity and increase the EMI shielding effectiveness of gypsum [14]. With 3 mm long isotropic pitch-based carbon fibers in the amount 2.0% of the weight of the gypsum (α-CaSO$_4$·0.5H$_2$O, 96%) or 1.7 vol.% of the gypsum plaster or 27 kg fibers/m^3 gypsum plaster, the electrical resistivity is decreased from 3.4×10^6 to 6.0×10^2 Ω.cm [14]. The minimum amount of carbon fibers for the resistivity to decrease is 0.5% of the weight of the gypsum [14]. With fibers in the amount of 2.0% of the weight of the gypsum, the EMI shielding effectiveness at 1.0 GHz is increased from 0.2 to 18 dB [14].

Carbon fibers are bound to increase the thermal conductivity of concretes and gypsums. An increased thermal conductivity in concretes is attractive for heated bridges and floorings of buildings with radiant heating. An increased thermal conductivity in gypsums is attractive for plaster moldings that allow a high rate of casting solidification [15].

Carbon Fiber Reinforced Glasses

Carbon fiber reinforced glasses are useful for space structural applications, such as mirror back structures and supports, booms and antenna structures. In low earth orbit these structures experience a temperature range of −100–80°C, so they need an improved thermal conductivity and a reduced coefficient of thermal expansion. In addition, increased toughness, strength, and modulus are desirable. Due to the environment degradation resistance of carbon fiber reinforced glasses, they are also potentially useful for gas turbine engine components. Additional attractions are low friction, high wear resistance, and low density. However, the high temperature resistance of these composites is limited by the tendency of the carbon fibers to be oxidized in air above ∼ 500°C. Exposure of carbon fiber reinforced alkali borosilicate (ABS) glasses to 400°C in air for up to 72 h. has no effect on the room temperature properties of the composites [15].

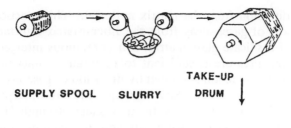

Figure 9.8   The slurry infiltration process for fabrication of continuous fiber ceramic-matrix composites. From Ref. 21. (Reprinted by permission of Kluwer Academic Publishers.)

The glass matrices used for carbon fiber reinforced glasses include borosilicate glasses (e.g. Pyrex), aluminosilicate glasses, soda lime glasses and fused quartz. Moreover, a lithia aluminosilicate glass-ceramic [16] and a $CaO–MgO–Al_2O_3–SiO_2$ glass-ceramic [17] have been used.

The carbon fibers can be short or continuous. In both cases, the composites can be formed by viscous glass consolidation, i.e., either hot pressing a mixture of carbon fibers and glass powder [18,19], or winding glass-impregnated continuous carbon fibers under tension above the annealing temperature of the glass [20].

Short ($\leqslant 3$ mm) carbon fiber borosilicate glass composites are prepared by hot pressing, in vacuum or argon, an isopropyl alcohol slurry of fibers and Pyrex powder ($< 50 \mu$m particle size) at 700–1 000°C (depending on the fiber content, which ranges from 10 to 40 vol.%) and 6.9 MPa [18]. Fibers longer than 3 mm are more difficult to disperse in the slurry. The resulting composite has the fibers two-dimensionally random, as some alignment occurs during pressing.

Continuous fiber composites are made by allowing fibers from a spool to pass through a glass powder slurry (containing water and a water-soluble acrylic binder), winding the slurry-impregnated fibers onto the sides on a hexagonal prism (mandrel or take-up drum), cutting up the flat unidirectional tapes from the mandrel, stacking up the pieces (plies) in a proper orientation, burning out the stack to remove the binder, and hot pressing the stack at a temperature above the working temperature of the glass, as illustrated in Figure 9.8 [21]. This process is known as slurry infiltration or viscous glass consolidation. During hot pressing, the glass must flow into the space between adjacent fibers. Since glass does not wet carbon, a sufficient pressure is necessary [19].

During hot pressing, a reaction that forms CO, CO_2, and SiO gases, together with solid carbide or oxycarbide, tends to occur. For the case of Pyrex glass as the matrix, the fiber–matrix interface layer has been shown to be a sodium-rich silicon oxycarbide of thickness about 1 000 Å [22]. (The sodium results from diffusion from the matrix material.) This reaction can be limited by the transport of gases away from the fiber–matrix interface. For example, for the case of a silicate glass matrix, if a continuous interphase of SiC forms and the external pressure is sufficient to maintain contact between the fibers and the matrix, the reaction is limited by diffusion of gaseous reaction products and by diffusion of C and/or SiO_2 through the SiC interphase. The consequence is an adherent interface with high shear strength [19]. On the other hand, if the external pressure is not sufficient to maintain contact between the fibers and the matrix, a poorly adherent interface is the consequence [19]. The doping of the glass with an appropriate oxide can control the extent of the interfacial reaction [19,23].

Viscous glass consolidation can also be carried out without hot pressing, but by winding under tension (e.g., about 15 ksi or 100 MPa) a glass powder-impregnated continuous fiber tow onto a collection mandrel at a temperature above the annealing temperature of the glass [20].

Other than viscous glass consolidation, another method for forming short or continuous carbon fiber glass-matrix composite is the sol–gel method, i.e., infiltration by a sol and subsequent sintering [24], using metal alkoxides and/or metal salts as precursors and using an acid or base catalyst to promote hydrolysis and polymerization [25]. Composite fabrication consists of two steps: (1) the preparation of fiber-gel prepregs and (2) thermal treatment and densification by hot pressing [26]. In the first step, the sol-impregnated prepregs are allowed to gel, dried at room temperature for a day, dried at 50°C for a day, then heat-treated at 300–400°C for 3 h. In the second step, hot pressing is performed in a nitrogen atmosphere with a pressure of 10 MPa [26]. For the case of a borosilicate glass, the hot pressing temperature is between 900 and 1 200°C [25]. The sol–gel method is advantageous in that homogenization of components during impregnation can readily be achieved and sintering temperatures are substantially lowered because of the smaller particle size compared to the slurry infiltration method. The easier impregnation or infiltration decreases the tendency for the particles to damage the fibers. However, the sol–gel method has the disadvantage of excessive shrinkage during subsequent heat treatment [21].

The slurry infiltration and sol–gel methods can be combined by using particle-filled sols. The particle filling helps to reduce the shrinkage [21].

Yet another method is melt infiltration, which requires heating the glass at a temperature much above the softening temperature in order for the glass to infiltrate the fiber preform. This high temperature may lead to chemical reaction between the fibers and the matrix [21].

The reaction between SiO_2 and carbon fibers is of the form:

$$3C + SiO_2 \rightarrow SiC + 2CO.$$

To reduce the extent of this reaction, SiC-coated carbon fibers are used for SiO_2-matrix composites.

The carbon fibers have the following effects on the glass: (1) increasing

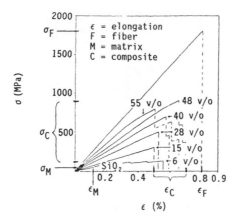

Figure 9.9 Tensile stress–strain curves of silica glass reinforced with different volume fractions of continuous SiC-coated carbon fibers. The stress is parallel to the fibers. σ_M = matrix strength, σ_F = fiber strength, σ_C = composite strength, ε_M = matrix strain at failure, ε_F = fiber strain at failure, ε_C = composite strain at failure. From Ref. 28.

the toughness (and, in some cases, the strength as well), (2) decreasing the coefficient of thermal expansion, and (3) increasing the thermal conductivity.

The increase in toughness is present in glass with short and randomly oriented fibers, as well as glass with continuous fibers. However, the strength is usually decreased by the fibers in the former and is usually increased by the fibers in the latter. Even for continuous fibers, the strengthening is limited by the widespread matrix cracking (both transverse and longitudinal) [27].

Figure 9.9 shows the tensile stress–strain curves of silica glass containing various volume fractions of continuous SiC-coated carbon fibers, compared to those of SiO_2 and carbon fibers [28]. Tensile testing along the fiber direction does not give results that depend on the fiber–matrix bonding, but flexural testing does. Thus, flexural testing is more revealing than tensile testing along the fiber direction. Figure 9.10 shows the load–deflection curves during flexural testing of short carbon fiber reinforced Pyrex and plain Pyrex [18]. The fibers cause a tail in the curve (i.e., increased toughness), even though they reduce the flexural strength. This behavior is an indication of the poor bonding between the fibers and the matrix. However, by aligning the short fibers [18] or using continuous fibers [19], the flexural strength of the composite can be higher than the matrix value. Figure 9.11 shows the effect of continuous carbon fibers on the flexural strength of glass-ceramic at different temperatures in a nitrogen atmosphere (for avoiding fiber oxidation) [24]. The flexural strength is increased by about 500%.

The high-temperature strength of carbon fiber reinforced glasses in air is limited by the oxidation of the carbon fibers. This oxidation cannot be prevented by the glass matrix [29]. In an inert atmosphere, the high-

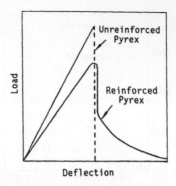

Figure 9.10 Load–deflection curves for unreinforced Pyrex and short carbon fiber reinforced Pyrex. From Ref. 18. (By permission of Chapman & Hall, London.)

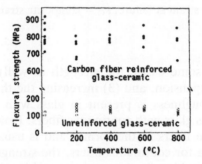

Figure 9.11 Flexural strength of unreinforced and continuous carbon fiber reinforced glass-ceramic in argon at five temperatures. From Ref. 24. (Reprinted by courtesy of Springer-Verlag, Heidelberg.)

temperature strength is limited by the softening of the glass matrix. Figure 9.12 shows the flexural strength of continuous carbon fiber (50–60 vol.%) reinforced borosilicate glass in argon as a function of temperature. The strength increases at 500°C because the glass matrix begins to soften, enhancing the ability to redistribute loads in the specimen. At 700°C, the strength is greatly decreased and extensive deformation occurs without any fracture, because the glass matrix has become very soft [29]. For glass-ceramic matrices, the strength is retained to higher temperatures, as shown in Figure 9.11.

The toughness of borosilicate glass with continuous carbon fibers is retained at cryogenic temperatures, such as −150°C [30].

The shear modulus of unidirectional continuous carbon fiber reinforced borosilicate glass is higher than that of the corresponding angle-ply (multiaxial) composites. This is attributed to more extensive matrix microcracking in the multiaxially reinforced composite. Cooling from the fabrication temperatures

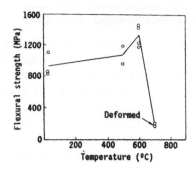

Figure 9.12   Flexural strength of continuous carbon fiber reinforced borosilicate glass in argon as a function of temperature. From Ref. 29.

results in residual stresses due to the difference in thermal expansion between the fiber and the matrix. In angle-ply composites, additional stresses are generated due to ply anisotropy in thermal expansion [31].

The in-plane coefficient of thermal expansion (-20–$80°C$) of continuous carbon fiber reinforced borosilicate glass decreases with increasing fiber content, such that its values are $0.36 \times 10^{-6}/°C$ at 36 vol.% fibers and $-0.23 \times 10^{-6}/°C$ at 58 vol.% fibers [19].

Hysteresis (difference between heating and cooling) in the curve of thermal expansion versus temperature occurs in two-dimensional continuous carbon fiber reinforced glasses, such as ABS. Much less hysteresis occurs in one-dimensional composites. No hysteresis occurs in the neat glass or the fibers themselves. The hysteresis is attributed to the slippage of the fibers at the fiber–matrix interface, which is weak. The transverse plies in the two-dimensional composites are believed to cause unequal resistance to motion during the heating and cooling portions of each thermal cycle. In contrast, the fibers in the one-dimensional composites can slide with equal resistance during heating and cooling. Multiple thermal cycling of the composite and strengthening of the fiber–matrix bond greatly reduce the hysteresis, though the bond strengthening also weakens and embrittles the composite. The reduction of the hysteresis is important because the thermal strains of mirrors (with the composites as mirror backs) must be predictable as the mirrors go through each thermal cycle, so that appropriate corrections can be made to insure positional accuracy [15].

In addition to decreasing the coefficient of thermal expansion, the carbon fibers increase the thermal conductivity, thus enhancing the thermal stability of fused quartz mirrors for space telescopes [32]. However, the thermal diffusivity is irreversible on heating and cooling, due to a thermal-history-dependent thermal barrier resistance at the fiber–matrix interface [16].

Other Carbon Fiber Reinforced Ceramics

MgO, Al_2O_3, SiC, Si_3N_4, mullite, β'-sialon, and ZrO_2 are other ceramics that have been used as matrices for carbon fiber composites. These composites are made by a number of methods, including (1) sintering the ceramic matrix powder and the carbon fibers, (2) sintering the ceramic precursor and the carbon fibers, (3) polymer infiltration and pyrolysis, (4) infiltrating the carbon fiber preform with a ceramic sol then subsequently gelling and sintering, (5) infiltrating the carbon fiber preform with a molten solution (e.g., Si) which will react with the carbon fibers to form the desired ceramic (e.g., SiC), and (6) chemical vapor infiltration (CVI) of the carbon fiber preform.

In Method 1, a slurry containing the ceramic matrix powder (MgO powder in the case of an MgO-matrix composite, Al_2O_3 powder in the case of an Al_2O_3-matrix composite, etc.) and a vehicle (e.g., isopropyl alcohol) is combined with carbon fibers (short or continuous) then sintered. The sintering may be with or without pressure. For example, MgO-matrix composites containing 5–40 vol.% carbon fibers are prepared by hot pressing an isopropyl alcohol slurry containing MgO powder ($< 1 \mu m$ particle size) and short ($\leq 3 mm$ long) carbon fibers at $1\,000°C$ and $6.9\,MPa$, followed by firing at $1\,200°C$ [18]. Al_2O_3-matrix composites containing 10–30 vol.% carbon fibers are prepared by hot pressing an isopropyl alcohol slurry containing Al_2O_3 powder ($\sim 0.3 \mu m$ particle size) and short ($\leq 3 mm$) carbon fibers at $1\,400°C$ and $20.7\,MPa$ [18]. The flexural toughness is increased while the flexural strength is decreased by the fiber addition. For example, for an Al_2O_3-matrix composite containing 20 vol.% fibers, the flexural toughness is increased by 75% while the flexural strength is decreased by 72%; for an MgO-matrix composite containing 20 vol.% fibers, the flexural toughness is increased by 800% while the flexural strength is decreased by 90% [18].

Method 2 is similar to Method 1 except that a ceramic precursor is used instead of the ceramic matrix powder. During sintering, the precursor is converted to the desired ceramic matrix phase. The β'-sialon $Si_{6.25}Al_{0.75}O_{0.75}N_{7.25}$ is a matrix that can be formed upon reaction among α-Si_3N_4, Al_2O_3, and AlN [33]. The Si_3N_4 (reaction-bonded silicon nitride, abbreviated RBCN) is a matrix that can be formed upon reaction of silicon powder with N_2 gas at $1\,200$–$1\,400°C$ [33]. Carbon fibers are particularly suitable for Si_3N_4-matrix composites because of their stability at high temperatures in a nitrogen atmosphere [34,35]. The Al_2O_3 is a matrix that can be formed upon oxidation of aluminum, using the LANXIDE process, which involves infiltrating a fiber preform with liquid aluminum and simultaneously oxidizing the aluminum matrix by flowing oxygen. The AlN is a matrix which can be formed also by the LANXIDE process, such that nitrogen is used instead of oxygen as the reactant gas. The LANXIDE process suffers from the presence of residual aluminum in the resulting composite. In general, the advantage of Method 2 is that there is no shrinkage and no liquid-phase

Table 9.3 Polymers for making ceramics by pyrolysis. From Ref. 36.

Polymer(s)	Resulting ceramic
Poly(silazanes)	Si_3N_4
Poly(silazanes)	Si–C–N
Polytitanocarbosilane	Si–Ti–C
Poly(carbosilanes)	SiC, Si–C–N
Polysilastyrene	SiC
Carboranesiloxane	$SiC–B_4C$
Polyphenylborazole	BN

sintering aids, which may limit the high-temperature properties of the resulting composite [21].

Method 3, polymer infiltration and pyrolysis, involves infiltrating the fiber preform with a polymer, such as an organometallic, and subsequent pyrolysis to convert the polymer into a ceramic. The requirements of the polymer are: (1) high ceramic yield, (2) low viscosity for ease of infiltration, and (3) ability to wet the fibers. A disadvantage of this technique is the high shrinkage during pyrolysis, as this shrinkage necessitates successive infiltration and pyrolysis operations. After pyrolysis, heat treatment at a higher temperature is carried out to improve the crystallinity of the ceramic. Method 3 is most commonly used for SiC-matrix composites, for which polycarbosilane (an organometallic) can be used as the polymer, which is pyrolyzed at 500–1 000°C and subsequently heat-treated at temperatures up to 1 500°C [21]. Table 9.3 [36] lists the polymers used for making various ceramics by pyrolysis.

Method 4 involves infiltrating the carbon fiber preform (containing short or continuous carbon fibers) with a ceramic sol (alumina sol in the case of an Al_2O_3-matrix composite, a mullite sol in the case of a mullite-matrix composite, etc.) and subsequent gelling of the fiber/sol mixture, followed by drying and sintering. A method for preparing an alumina sol involves using boehmite gel powder and 0.5 μm α-alumina particles, such that 80% of the oxide alumina results from the α-alumina particles and the remaining 20% from the boehmite [37]. The further addition of a silica sol to the alumina sol yields a mullite sol [37]. After infiltrating the carbon fiber preform with a sol, gelling occurs in about 20 min. After that, drying takes place at 50°C for 20 h. Subsequently hot pressing is carried out at 1 200–1 300°C for 1 h. at a pressure of 20 MPa [37]. The sol–gel method suffers from excessive shrinkage, but it is suitable for the production of multicomponent matrices [21].

Method 5 uses an infiltrant which is molten at the infiltration temperature, reacts with the carbon fibers, and wicks into the fiber preform. For a SiC-matrix composite, the infiltrant can be a mixture of silicon and boron powders such that boron is in an amount of about 3% by weight of silicon [38].

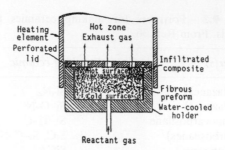

Figure 9.13 The thermal and pressure gradient method of chemical vapor infiltration. From Ref. 21. (Reprinted by permission of Kluwer Academic Publishers.)

The powder mixture is heated in a vacuum to about 1 450°C to form a saturated solution together with a finely divided precipitate of a compound of boron and silicon [38]. The infiltration of the fiber preforms occurs at about 1 420°C, at which the infiltrant is fluid [38]. The resulting matrix is predominantly SiC and/or boron-containing SiC. A variation of this method involves converting the carbon matrix of a carbon–carbon composite to SiC by reaction with liquid silicon. This method requires that the fibers be only slightly attacked by the liquid silicon. Its main disadvantage is the residual unreacted silicon, which limits the high-temperature properties above 1 410°C (the melting point of silicon) [21].

Method 6 is chemical vapor infiltration (CVI). For the case of a SiC-matrix composite, the vapor can be methyltrichlorosilane (CH_3SiCl_3) in an H_2 carrier gas at a temperature of about 1 000°C [39]. Zirconia (ZrO_2) is another matrix that can be formed by CVI [40]. This method is most amenable to continuous fibers, but it is a slow process (taking weeks or months) [41]. The slowness is necessary in order to attain a uniform deposition rate throughout the material. Otherwise, deposition is mostly at the outer regions of the sample, which is thus closed off. In order for the process to be slow, the temperature and pressure are kept low. Only composites less than about 3 mm thick can be produced without significant density gradients. The infiltration time can be considerably reduced (to < 24 h.) by using thermal and pressure gradients. In the thermal and pressure gradient method of CVI, the top of the fiber preform is exposed to the hot zone of the furnace while the preform is contained in a water-cooled graphite holder, which cools the reactant gas inlet (at the bottom of the holder) and the sides and bottom of the preform, as illustrated in Figure 9.13. The reactant gas is forced under pressure into the cooled portion of the preform, but initially does not react because of the low temperature. Gases pass from the cooled portion of the preform into the hot portion, where deposition begins. As deposition occurs, the hot zone becomes denser, so the deposition front progressively moves from the top to the bottom of the preform [21]. The rate of deposition also depends on the fiber. It increases with increasing active surface area of the fiber. (The active sites are the defects in the aromatic basal planes, as detected by oxygen chemisorption.) The active surface area may be increased by oxidation of the fiber in air [42].

The choice of a ceramic matrix depends largely on the high-temperature stability of the ceramic in the presence of carbon fibers. Carbide matrices (e.g., SiC, TiC) are attractive because of their chemical compatibility with carbon. Among the noncarbide matrices, AlN and Al_2O_3 are more stable compared to TiN, Si_3N_4, TiO_2 and SiO_2, according to thermodynamic calculations, which consider the following reactions between carbon and the various ceramic phases [43].

$$3C + Si_3N_4 \rightarrow 3SiC + 2N_2$$

$$2C + 2TiN \rightarrow 2TiC + N_2$$

$$3C + 4AlN \rightarrow Al_4C_3 + 2N_2$$

$$3C + TiO_2 \rightarrow TiC + 2CO$$

$$3C + SiO_2 \rightarrow SiC + 2CO$$

$$9C + 2Al_2O_3 \rightarrow Al_4C_3 + 6CO$$

All C–nitride and C–oxide reactions involve the release of gaseous N_2 and CO, respectively, and are therefore dependent on the external atmosphere, the gas pressure, and the porosity of the composite materials [43]. The thermodynamic calculations show that stable C–ceramic matrix composite materials may be obtained in inert atmospheres with AlN and Al_2O_3 up to approximately 1 600°C [43]. However, the C–Al_2O_3 reaction shown above occurs in an inert gas at temperatures below 1 600°C when the CO pressure is low enough; this CO pressure increases as the activity of oxygen in Al_2O_3 increases. To inhibit the reaction between the carbon fibers and the ceramic matrix, the carbon fibers may be coated with a less reactive ceramic material, such as SiC, though spontaneous delamination of the coating is a problem [44]. This reactivity between the fibers and the matrix limits the sintering temperature as well as the use temperature.

Besides the reactivity between the fibers and the matrix, the choice of a ceramic matrix also depends on the thermal conductivity, as a high thermal conductivity is desirable for numerous applications, such as in space structures and electronic packaging. Different ceramic materials can differ greatly in thermal conductivity. For example, AlN is more conductive than SiC, which is in turn more conductive than Al_2O_3.

The choice of a ceramic matrix may also depend on the presence of a polymorphic transformation, which may be useful for increasing the toughness of the ceramic matrix. Zirconia (ZrO_2) is a ceramic material that exhibits such a transformation.

From a practical point of view, the choice of a ceramic matrix depends largely on the possibility of using a convenient method to prepare composites that are sufficiently low in porosity. Certain methods are more suitable for

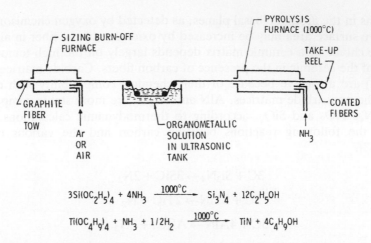

$$3Si(OC_2H_5)_4 + 4NH_3 \xrightarrow{1000°C} Si_3N_4 + 12C_2H_5OH$$

$$Ti(OC_4H_9)_4 + NH_3 + 1/2H_2 \xrightarrow{1000°C} TiN + 4C_4H_9OH$$

Figure 9.14 Process for coating carbon fibers with a nitride by using an organometallic solution. From Ref. 46. (Reprinted by courtesy of Marcel Dekker, Inc.)

short fibers than continuous fibers; certain methods are established only for several ceramic matrices.

A special class of ceramic-matrix composites is those with inorganic polymer matrices, such as phosphate binders (metal oxides plus phosphoric acid) and magnesium binders (magnesium chloride solutions). The chemical stability of carbon fibers in their presence makes these binders suitable for carbon fiber composites. An example of a phosphate binder is the copper phosphate binder ($CuO–H_3PO_4$ system), which forms mainly $Cu(H_2PO_4)_2$ after curing for 90–240 min., and mainly $CuHPO_4$ after curing for ≥ 1 day [45].

Control of the fiber–matrix bonding is important for ceramic-matrix composites. For this purpose, coatings can be applied on the carbon fibers. For example, Si_3N_4 coatings can be applied by reaction of an organometallic, namely $Si(OC_2H_5)_4$, with NH_3 at $1\,000°C$, and TiN coatings can be applied by reaction of an organometallic, namely $Ti(OC_4H_9)_4$, with H_2 at $1\,000°C$, as illustrated in Figure 9.14 [46]. A TiN coating has been used to reduce the fiber–matrix reaction in Si_3N_4-matrix carbon fiber composites [35]. SiC and TiC coatings have been used to reduce the reaction between Si and the carbon fibers in composites with a SiC matrix formed by reaction bonding (i.e., reaction between Si and C) [28]. An additional function of the coatings is to increase the oxidation resistance of the fibers [42].

References

1. P.-W. Chen and D.D.L. Chung, *Composites* **24**(1), 33–52 (1993).
2. X. Yang and D.D.L. Chung, *Composites* **23**(6), 453–460 (1992).
3. P.-W. Chen and D.D.L. Chung, *Smart Mater. Struct.* **2**, 22–30 (1993).
4. J.-M. Chiou, Q. Zheng, and D.D.L. Chung, *Composites* **20**(4), 379–381 (1989).

5. P.-W. Chen and D.D.L. Chung, *Ext. Abstr. Program—Bienn. Conf. Carbon,* **21**, 92–93 (1993).
6. Q. Zheng and D.D.L. Chung, *Cement and Concrete Research* **19**, 25–41 (1989).
7. S. Furukawa, Y. Tsuji, and S. Otani, in *Proc. 30th Jpn. Congr. Mater. Res.,* 1987, pp. 149–152.
8. S. Akihama, T. Suenaga, and T. Banno, *Int. J. of Cement Composites & Lightweight Concrete* **8**(1), 21–33 (1986).
9. S. Akihama, M. Kobayashi, T. Suenaga, H. Nakagawa, and K. Suzuki, *KICT Report 65*, Kajima Institute of Construction Technology, October 1986.
10. Y. Ohama, Y. Sato, and M. Endo, in *Proc. Asia–Pacific Concrete Technology Conf. 1986*, Institute for International Research, Singapore, 1986, pp. 5.1–5.8.
11. S.B. Park, P.I. Lee, and Y.S. Lim, *Cem. Concr. Res.* **21**(4), 589–600 (1991).
12. H. Hahne, S. Karl, and J. Worner, *Fiber Reinforced Concrete Properties and Applications, SP105*, American Concrete Institute, Detroit, MI, 1987, pp. 211–223.
13. R. Lankard and K. Newell, *Fiber Reinforced Concrete, SP81*, American Concrete Institute, Detroit, MI, 1984, pp. 287–306.
14. D.D.L. Chung and Q. Zheng, *Compos. Sci. Technol.* **36**, 1–6 (1989).
15. V.F. Janas, in *Proc. Int. SAMPE Symp. and Exhib., 33, Materials: Pathway to the Future*, edited by G. Carrillo, E.D. Newell, W.D. Brown, and P. Phelan, 1988, pp. 357–368.
16. D.P.H. Hasselman, *Therm. Conduct.* **19**, 383–402 (1988).
17. H.S. Kim, R.D. Rawlings, and P.S. Rogers, *Br. Ceram. Proc.*, **42**, 59–68 (1989).
18. R.A.J. Sambell, D.H. Bowen, and D.C. Phillips, *J. Mater. Sci.* **7**, 663–675 (1972).
19. W.K. Tredway, K.M. Prewo, and C.G. Pantano, *Carbon* **27**(5), 717–727 (1989).
20. R.A. Allaire, U.S. Patent 4,976,761 (1990).
21. B. Rand and R.J. Zeng, in *Carbon Fibers Filaments and Composites*, edited by J.L. Figueiredo, C.A. Bernardo, R.T.K. Baker, and K.J. Huttinger, Kluwer Academic, Dordrecht, 1990, pp. 367–398.
22. S.M. Bleay and V.D. Scott, *Microbeam Anal.* **26**, 234–236 (1991).
23. C.G. Pantano, G. Chen, and D. Qi, *Mater. Sci. Eng.* **A126**, 191–201 (1990).
24. R.A.J. Sambell, D.C. Phillips, and D.H. Bowen, *Carbon Fibres, Their Composites and Applications, Proc. of Int. Conf.*, The Plastics Institute, 1974, pp. 105–113.
25. V. Gunay, P.F. James, F.R. Jones, and J.E. Bailey, *Br. Ceram. Proc.*, **45**, 229–240 (1989).
26. V. Gunay, P.F. James, F.R. Jones, and J.E. Bailey, *Inst. Phys. Conf. Ser.*, Vol. 111, *New Materials and Their Applications 1990*. IOP, Bristol, U.K. and Philadelphia. 1990, pp. 217–226.
27. F.A. Habib, R.G. Cooke, and B. Harris, *Br. Ceram. Trans. J.* **89**, 115–124 (1990).
28. E. Fitzer, *High Temp.–High Pressures* **18**(5), 479–508 (1986).
29. K.M. Prewo and J.A. Batt, *J. Mater. Sci.* **23**, 523 (1988).
30. D.F. Hasson and S.G. Fishman, *Ceram. Eng. Sci. Proc.* **11**(9–10), 1639–1647 (1990).
31. V.C. Nardone and K.M. Prewo, *J. Mater. Sci.* **23**, 168–180 (1988).
32. Jiang Yasi, Jiang Shibin, Wang Huirong, Chen Menda, and Xie Ruibao, in *Proc. SPIE: Int. Soc. Opt. Eng., 1236 (Adv. Technol. Opt. Telesc. 4, Pt. 2)*, 1990, pp. 712–715.

33. R.M. Brown, H.J. Edrees, and A. Hendry, *Br. Ceram. Proc.*, **45**, 169–177 (1989).

34. D.B. Fischbach, in *Proc. Int. Carbon Conf., Baden-Baden, Germany*, 1986, pp. 719–721.

35. B. Saruhan and G. Ziegler, *Silic. Ind.* **55**(1–2), 29–32 (1991).

36. K.J. Wynne and R.W. Rice, *Ann. Rev. Mater. Sci. 14*, edited by R.A. Huggins, Annual Reviews, Palo Alto, CA, 1984, 297–334.

37. M. Chen, F.R. Jones, P.F. James, and J.E. Bailey, *Inst. Phys. Conf. Ser., Vol. 111, New Mater. and Their Applications 1990*, IOP, Bristol, U.K. and Philadelphia, 1990, pp. 227–237.

38. R.N. Singh and W.A. Morrison, U.S. Patent 4,944,904 (1990).

39. R. Taylor and V. Piddock, *Mater. Sci. Forum* **34–36**, 525–530 (1988).

40. J. Minet, F. Langlais, J.M. Quenisset, and R. Naslain, *J. Eur. Ceram. Soc.* **5**(6), 341–356 (1989).

41. S.C. Danforth, in *Proc. Int. Symp. Adv. Processing of Ceramic and Metal Matrix Composites, Halifax, 1989*, edited by H. Mostaghaci, Pergamon, New York, 1989, pp. 107–119.

42. P. Ehrburger and J. Lahaye, *High Temp.–High Pressures* **22**(3), 309–316 (1990).

43. P. Greil, *J. Eur. Ceram. Soc.* **6**(1), 53–64 (1990).

44. V. Gupta, A.S. Argon, H.S. Landis, and J.A. Cornie, *Ceram. Eng. Sci. Proc.* **9**(7–8), 985–992 (1988).

45. I.N. Ermolenko, I.P. Lyubliner, and N.V. Gulko, *Chemically Modified Carbon Fibers*, translated by E.P. Titovets, VCH, Weinheim, Germany, 1990.

46. H.A. Katzman, *Mater. Manufacturing Processes* **5**(1), 1–15 (1990).

CHAPTER **10**

Hybrid Composites

Hybrid composites usually refer to composites containing more than one type of filler and/or more than one type of matrix. They are commonly used for improving the properties and/or lowering the cost of conventional composites. The possibility of using more than one type of filler or more than one type of matrix should always be considered in composite design.

When more than one type of filler is used in a carbon fiber composite, the second type of filler may be a different type of fiber; it may even be whiskers or particles. When both fillers are fibers, one or both of the fillers can be short or continuous. The addition of a second type of fiber, such as glass or aramid fibers, can serve to increase the toughness of the carbon fiber composite. The two types of fibers may be in different laminae or the same lamina in the composite laminate. If they are in the same lamina, they may be cowoven, twisted, stapled, randomly mixed, or bound by a binder. When the second filler is particles, the particles are usually intermixed with the fibers within a lamina by incorporating the particles or the particle precursor in the matrix precursor prior to impregnation of the matrix precursor into the fiber preform. However, they are sometimes located instead between adjacent fiber laminae, such that the matrix permeates the space between adjacent fibers as well as the space between adjacent particles. The particulate filler can serve to enhance the mechanical or electrical properties of the composite.

When two types of matrix are used in a composite, one of two cases applies. In the first case, one type of matrix is in contact with the filler while the second type of matrix is not; in other words, the second type of matrix is just in contact with the first type of matrix. This structure is obtained by infiltrating the carbon fiber preform with the precursor of the first type of matrix, and subsequently with the precursor of the second type of matrix. For example, the first type of matrix can be a thermoset (a lower-viscosity resin) while the second type of matrix can be a thermoplast (a higher-viscosity resin); the first type of matrix can be carbon (which is not oxidation resistant) while the second type of matrix can be SiC (which is more oxidation resistant). The second type of matrix is the one that is at the exterior surface of the composite.

In the second case, the two types of matrix are intimately mixed in a random fashion so that the mixture serves as the matrix. An example is the use of carbon and SiC together as the matrix by pyrolyzing a mixture of a carbon precursor (e.g., pitch) and a SiC precursor (e.g., a polycarbosilane) [1]. The interest behind the use of SiC and C together as the matrix lies in oxidation protection of the carbon composite.

A special class of hybrid composites (not described above) is the sandwich composites. A sandwich composite has an interlayer (interleaf) between adjacent carbon fiber laminae in the laminate, though an interlayer does not necessarily reside between every pair of fiber laminae. The interlayer itself may be a composite or not a composite. An example of an interlayer that is itself a composite is a honeycomb, such as an aluminum honeycomb formed by brazing aluminum sheets together in such a way as to form a structure that looks like a honeycomb (akin to corrugated paper). Honeycombs are attractive for their very low density, which is due to their very high porosity. Examples of interlayers that by themselves are not composites include aluminum and polymers, which serve to improve the toughness and/or control the thermal expansion coefficient.

Composites with More Than One Type of Filler

Ductile fibers such as glass [2], aramid [3,4], and polyethylene [5] fibers are used together with carbon fibers to improve the breaking toughness of composites. The "hybrid" effect refers to the enhancement in the failure strain of the low-elongation phase (carbon fibers) when part of a hybrid composite [6]. However, an increase in failure strain is accompanied by a decrease in stiffness, which can be predicted by the rule of mixtures. In particular, polyethylene fibers are used with carbon fibers in an epoxy matrix for providing a composite material with a high damping ability [7].

Figure 10.1 shows the measured tensile stress–strain curve of a unidirectional hybrid composite consisting of carbon fibers and glass fibers, together with that calculated for the hypothetical case in which the two kinds of fibers present are not bonded to the polymer matrix. The breaking strain (ε_C) of the carbon fiber composite portion is smaller than that (ε_{CG}) of the hybrid composite. If there is no bonding between the carbon fibers and the glass fibers (via the polymer matrix), the carbon fiber composite portion will be broken at strain ε_C and stress σ_C. After a reduction in the load, only the glass fiber composite portion will withstand the load to strain ε_G, at which the composite fractures. On the other hand, if both carbon and glass fibers are sufficiently well bonded to the matrix, when the carbon fibers begin to be partially broken, the glass fibers situated near the breaking portions share the load, through shearing of the matrix, and cut carbon fibers are effective as reinforcing short fibers [2].

The most commonly used method to toughen carbon fiber epoxy-matrix composites is the incorporation of an elastomer phase in the glassy epoxy

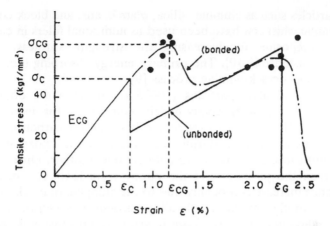

Figure 10.1 Measured tensile stress–strain curve (dotted curve) of a hybrid composite containing unidirectional carbon and glass fibers bonded in an epoxy matrix. The solid curve is one calculated for the case where the carbon and glass fibers are not bonded to the matrix. From Ref. 2. (By courtesy of Marcel Dekker Inc.)

matrix. The elastomer can be carboxyl-terminated butadiene–acrylonitrile (CTBN) or other nitrile rubbers added as a liquid to the epoxy resin. The rubber forms particles (about 5 μm in size) in the sea of the epoxy resin. Thus, the rubber particles constitute a filler, even though they are added as a liquid to the matrix resin. The fracture mechanism of the hybrid composite containing carbon fibers and rubber particles depends on the strain rate and temperature. At a high strain rate or a low temperature, the rubber particles are torn, because the fracture propagates through these particles; this results in a high strength. At a low strain rate or a high temperature, the rubber particles are not torn, because the fracture propagates around these particles; this results in a low strength [8]. Fracture mechanisms that might operate to enhance the impact resistance of the epoxy include crazing, shear band forming, and voiding for stress relaxation [9]. However, a large increase in resin toughness does not give a proportional increase in composite toughness, and the toughened resin may also result in a reduction in strength and modulus of the composite [9].

An alternative to toughening by the use of rubber particles is toughening by the use of thermoplastic particles, such as poly(vinylidene fluoride) (PVDF) and PEEK, in the amount of 5–15 wt.% of the epoxy resin. This alternative is attractive in that the impact energy of the carbon fiber composite is increased without sacrificing the composite modulus and strength. In contrast to the rubber particles, the thermoplastic PEEK particles do not improve the toughness of the neat resin, though they enhance the energy-dissipating capability of the fiber composites. This is because the thermoplastic particles promote multiple cracking and delamination modes [9].

Rigid particles such as alumina, silica, glass beads, and block copolymers, as well as ceramic whiskers, have been used as additional fillers in carbon fiber epoxy-matrix composites for increasing the strength, stiffness, toughness, and/or fatigue resistance [9,10]. The primary energy-absorbing mechanism for these composites is a crack-pinning process [9].

Ductile tin–lead alloy particles (20–25 μm in size) are effective as a second filler in carbon fiber epoxy-matrix composites for increasing their fatigue resistance. These particles are added between the carbon fiber prepreg layers and experience melting during the curing of the epoxy matrix. The melting does not cause the particles to connect one another, but enhances the bonding between the particles and the epoxy resin [11]. The origin of the fatigue resistance increase is probably due to the stopping of crack propagation from one ply to another by the ductile alloy particles between the plies.

Tin–lead alloy particles (20–25 μm in size) are effective as a second filler in polymer-matrix composites containing short metal-coated carbon fibers for greatly decreasing the electrical resistivity of the composite. By adding 2 vol.% tin–lead particles to 20 vol.% short nickel-coated carbon fiber filled polyether sulfone, the electrical resistivity was decreased by a factor of 2 000, while the electromagnetic interference shielding effectiveness at 1 GHz was increased from 19 to 45 dB. This effect is due to the melting of the tin–lead particles during composite fabrication and the affinity of the molten alloy to the metal-coated carbon fibers. This affinity causes the alloy to connect some of the carbon fibers, instead of remaining as discrete particles. The effect is much less if bare carbon fibers are used instead of metal-coated carbon fibers [12].

Silicon carbide whiskers of diameter 0.05–1.50 μm have been used as a second filler in carbon fiber (7 μm diameter) aluminum-matrix composites for improving the wear resistance (by acting as a barrier against slip of relatively larger fillers) [13] and for increasing the transverse strength [14]. The whiskers can be first attached to the surfaces of the carbon fibers prior to squeeze casting [14]. They can alternatively be mixed with the carbon fibers prior to composite fabrication by powder metallurgy [13]. To further improve the wear resistance, SiC particles (1.5–5 μm in diameter) may be added as a third filler, resulting in a hybrid composite containing 10 vol.% SiC particles, 5 vol.% SiC whiskers, and 4 vol.% carbon fibers [13].

Micrometer-sized SiC particles are used as a second filler in carbon–carbon composites by adding the SiC particles to the carbon matrix precursor (e.g., a thermoset resin) prior to impregnating the carbon matrix precursor into the carbon fiber preform. The SiC particles serve to decrease the shrinkage of the matrix during pyrolysis, so that only a single impregnation cycle is necessary. The SiC particles also serve to control the rheology of the carbon matrix precursor during processing. The resulting hybrid composite contains 40 vol.% fibers, 40–50 vol.% SiC, and 10–20 vol.% carbon [15].

A hybrid fiber is a fiber that consists of two or more components which are artificially put together. An example of a hybrid fiber is a PAN core surrounded by a vapor grown carbon fiber sheath [16] for the purpose of achieving high strength (with the core) and low electrical resistivity (with the sheath). Another example is a carbon fiber with a superconducting niobium carbonitride coating for obtaining a superconducting fiber [17].

Composites with More Than One Type of Matrix

Carbon–carbon composites contain carbon fibers in a carbon matrix that is often not fully dense. The filling of the pores in a carbon–carbon composite with SiC, TiC, or other ceramic materials that are more oxidation resistant than carbon yields a hybrid composite with more than one type of matrix.

Chemical vapor infiltration (CVI) has been used for the deposition of SiC [18] and TiC [19]. In particular, the thermal gradient CVI method has been used for the codeposition of C and SiC at 1 100–1 400°C, with propane (C_3H_8) and methyltrichlorosilane (CH_3SiCl_3) as the sources of carbon and silicon (carbide-forming element) respectively. Hybrid composites containing 12–29 wt.% SiC have been prepared, such that a CH_3SiCl_3/N_2 ratio (N_2 being the carrier gas) increase gives rise to a SiC weight fraction increase. At low deposition temperatures (1 100–1 200°C), SiC codeposition does not significantly affect the microstructure of the carbon matrix. Since SiC is stiffer than carbon, the mechanical properties (e.g., Young's modulus) of the composites are improved by SiC codeposition. However, at high deposition temperatures (1 300–1 400°C), due to the increase of the degree of crystalline order of the carbon matrix by the SiC codeposition, a matrix of columnar structure and large crystallite size is formed, resulting in no improvement of the mechanical properties by the SiC codeposition [20].

Impregnation of a SiC precursor (polycarbosilane) into a carbon–carbon composite has also been achieved by the use of supercritical fluids, which typically exhibit densities approaching that of their liquid phase but have no surface tension and very low viscosity. The supercritical fluids dissolve, transport, and precipitate ceramic precursors within the pores of the carbon–carbon composite [21,22].

Titanium carbide is attractive because of its high thermal stability compared to SiC. Furthermore, TiC has a CVI infiltration capability much higher than those of HfC and TaC [19].

Due to the lower viscosity of epoxy resins than thermoplastics and the better fiber–matrix adhesion for epoxy than thermoplastic matrices, carbon fiber preforms may be partially impregnated with epoxy and subsequently fully impregnated with a thermoplastic. This results in a hybrid composite with more than one matrix.

Sandwich Composites

A thin and ductile polymer interleaf is placed between carbon fiber prepreg layers in order to increase (as much as doubling) the toughness of the composite by toughening the interface between the laminae in the laminate.

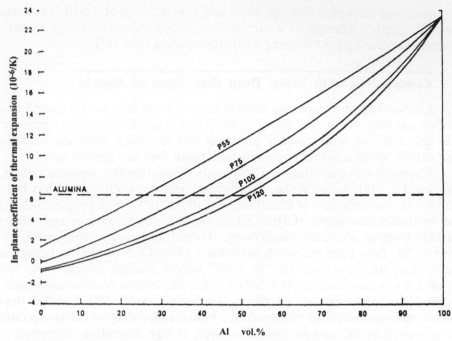

Figure 10.2 Calculated in-plane coefficient of thermal expansion of hybrid composites containing unidirectional carbon fiber epoxy-matrix composite layers alternating with aluminum sheets. From Ref. 26. (Reprinted by permission of the Society for the Advancement of Material and Process Engineering.)

The interleaf is formulated so that it remains a discrete well-bonded void-free layer after it is cocured with the matrix resin. It may be twenty times thicker than the resin between plies in a conventional laminate. The interleaf also serves to eliminate stress concentrations which otherwise would produce premature matrix failure [23].

The interleaf increases the area in which the contact force between an impactor and a laminate is distributed. The closer the interleaf is placed relative to the impact site, the larger is this contact area. In this way, the transverse shear stress concentration, and consequently matrix cracking, is reduced in the laminae above the interleaf. However, the interleaf does not affect the matrix cracking in the laminae beneath it. An optimum design involves placing the interleaf below the impacted face at a distance equal to the size of the contact area. If the interleaf is placed too far down, its effectiveness in increasing the contact area is diminished [24].

Instead of a polymer interleaf, an aluminum interleaf is used to increase the fatigue resistance of fiber epoxy-matrix composites. In this sandwich composite, an aluminum sheet is placed between adjacent fiber prepreg layers. Aluminum is chosen because of its ductility and low density (for aerospace use). These composites are known as Arall, as well as Fiber Metal Laminates [25].

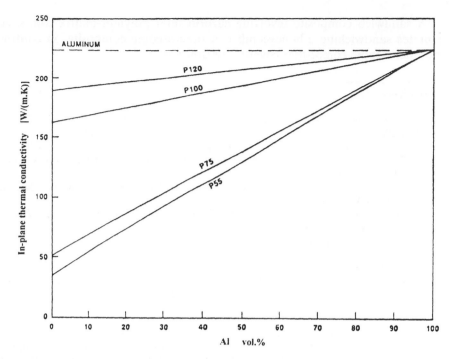

Figure 10.3 Calculated in-plane thermal conductivity of hybrid composites containing unidirectional carbon fiber epoxy-matrix composite layers alternating with aluminum sheets. From Ref. 26. (Reprinted by permission of the Society for the Advancement of Material and Process Engineering.)

A second function of the aluminum interleaf is to increase the coefficient of thermal expansion (CTE) of the carbon fiber composite (used as heat sinks) so as to match the coefficients of ceramics (e.g., alumina) used in electronic packaging. Figures 10.2 and 10.3 show the in-plane CTE and in-plane thermal conductivity calculated for composites in which the carbon fiber epoxy-matrix composite portion contains 60 vol.% unidirectional carbon fibers. To achieve a CTE of 6×10^{-6}/°C with a sandwich composite containing aluminum and a P-100 carbon fiber epoxy-matrix composite, an aluminum content of 45 vol.% is required (Figure 10.2). The corresponding thermal conductivity is about 109 Btu/h.-ft.-°F or 189 W/m/K (Figure 10.3) [26].

The high strength, low electrical resistivity, and low thermal expansion coefficient of carbon fibers are exploited in a sandwich composite in which a high-T_c ceramic superconductor is sandwiched by a tin-matrix unidirectional carbon fiber composite fabricated by squeeze casting. Tin (preferably 96Sn–4Ag) is used as the matrix because of the possibility of diffusion bonding it to the superconductor at a temperature low enough to avoid degradation of the superconducting properties of the superconductor. The sandwiching enables the superconductor to be packaged for mechanical durability, as the superconductor by itself is mechanically weak, especially under tension [27].

A sandwich composite involving carbon fiber polymer-matrix composite face plates sandwiching a honeycomb is a macroscopic composite, in contrast to the microscopic composites covered by this book.

References

1. K. Kawamura, M. Ono, and K. Okazaki, *Carbon* **30**(3), 429–434 (1992).
2. *Adhesion and Bonding in Composites*, edited by R. Yosomiya, K. Morimoto, A. Nakajima, Y. Ikada, and T. Suzuki, Marcel Dekker, New York, 1990, pp. 257–281. (Chapter on Interfacial Effect of Carbon-Fiber-Reinforced Composite Material.)
3. A. Haque, L. Moorehead, D.P. Zadoo and S. Jeelani, *J. Mater. Sci.* **25**, 4639–4643 (1990).
4. A. Haque, C.W. Copeland, D.P. Zadoo, and S. Jeelani, *J. Reinf. Plastics Compos.* **9**, 602–613 (1990).
5. A.A.J.M. Peijs, R.W. Venderbosch, and P.J. Lemstra, *Composites (Guildford, U.K.)* **21**(6), 522–530 (1990).
6. G. Kretsis, F. Matthews, J. Morton, and G. Davies, in *Proc. 3rd Eur. Conf. Compos. Mater., Dev. Sci. Technol. Compos. Mater.*, edited by A.R. Bunsell, P. Lamicq, and A. Massiah, Elsevier, London, 1989, pp. 671–675.
7. P. Raju Mantena and R.F. Gibson, in *Proc. 22nd Int. SAMPE Tech. Conf.*, 1990, pp. 370–382.
8. K. Nakao and Y. Yamashita, in *Composites '86: Recent Advances in Japan and the United States, Proc. Japan–U.S. CCM-III*, edited by K. Kawata, S. Umekawa, and A. Kobayashi, Jpn. Soc. Compos. Mater. Tokyo, 1986, pp. 743–750.
9. B.Z. Jang, J.Y. Liau, L.R. Hwang, and W.K. Shih, *J. Reinf. Plastics Compos.* **9**, 314–321 (1990).
10. M. Narkis and E.J.H. Chen, *SAMPE J.* **26**(3), 11–15 (1990).
11. S. Fang and D.D.L. Chung, *Composites (Guildford, U.K.)* **21**(5), 419–424 (1990); D.D.L. Chung, U.S. Patent 5,091,242 (1992).
12. L. Li and D.D.L. Chung, in *Proc. 6th Int. SAMPE Electron. Conf.*, 1992, pp. 652–658.
13. T.T. Long, T. Nishimura, T. Aisaka, and M. Morita, *Mater. Trans., JIM*, **32**(2), 181–188 (1991).
14. S. Towata, H. Ikuno, and S. Yamada, in *Proc. 6th ICCM Conf. Compos. Mater., 2nd Eur. Conf. Compos. Mater.*, Vol. 2, edited by F.L. Matthews, N.C.R. Buskell, J.M. Hodgkinson, and J. Morton, Elsevier, London, 1987, pp. 2.412–2.421.
15. F.I. Hurwitz, in *Proc. Natl. SAMPE Symp. and Exhibit.* 30, 1985, pp. 1375–1386.
16. J.R. Gaier, M.L. Lake, A. Moinuddin, and M. Marabito, *Carbon* **30**(3), 345–349 (1992).
17. M. Dietrich, E. Fitzer, and T. Stumm, *High Temp.–High Pressures* **19**(1), 89–94 (1987).
18. G.A. Green and F.J. Tribe, *Lubr. Eng.* **44**(8), 666–673 (1988).
19. J.Y. Rossignol, F. Langlais, and R. Naslain, *Proc. Electrochem. Soc.* **84**(6), 596–614 (1984).
20. H.-S. Park, D.-W. Kweon, and J.-Y. Lee, *Carbon* **30**(6), 939–948 (1992).
21. R.A. Wagner, V.J. Krukonis, and M.P. Coffey, *Ceram. Eng. Sci. Proc.* **9**(7–8), 957–964 (1988).
22. R.A. Wagner, V.J. Krukonis, and M.P. Coffey, in *Mater. Res. Soc. Symp. Proc., Vol. 121*, 1988, pp. 711–716.

23. R.B. Krieger, Jr., in *Carbon Fibers*, edited by The Plastics and Rubber Inst., London, England, Noyes, Park Ridge, NJ, 1986, pp. 146–157.
24. S. Rechak and C.T. Sun, *J. Reinf. Plastics Compos.* **9**, 569–582 (1990).
25. J.W. Gunnink and L.B. Vogelesang, in *Proc. Int. SAMPE Symp. and Exhib.*, *35*, *Advanced Materials: Challenge Next Decade*, edited by G. Janicki, V. Bailey, and H. Schjelderup, 1990, pp. 1708–1721.
26. K.A. Schmidt and C. Zweben, in *Proc. Int. SAMPE Electron. Conf.*, *3*, *Electron. Mater. Processes*, 1989, pp. 181–190.
27. C.T. Ho and D.D.L. Chung, *J. Mater. Res.* **4**(6), 1339–1346 (1989).

Index

Printed and bound by CPI Group (UK) Ltd, Croydon, CR0 4YY

03/10/2024

01040434-0002